BUILT-INS
AND STORAGE

FROM THE EDITORS OF **Fine Homebuilding**®

The Taunton Press

The Taunton Press

Inspiration for hands-on living®

The Taunton Press, Inc., 63 South Main Street, PO Box 5506, Newtown, CT 06470-5506

e-mail: tp@taunton.com

Distributed by Publishers Group West

Jacket/Cover Design: Cathy Cassidy

Interior Design: Cathy Cassidy

Layout: Carol Petro

Front Cover Photographer: Roe A. Osborne, courtesy *Fine Homebuilding,* © The Taunton Press, Inc.

Back Cover Photographers: (clockwise from top left) © Jim Tolpin; Charles Miller, courtesy *Fine Homebuilding,* © The Taunton Press, Inc.; Scott Gibson, courtesy *Fine Homebuilding,* © The Taunton Press, Inc.; Roe A. Osborn, courtesy *Fine Homebuilding,* © The Taunton Press, Inc.

Taunton's For Pros By Pros® and Fine Homebuilding® are trademarks of The Taunton Press, Inc., registered in the U.S. Patent and Trademark Office.

Library of Congress Cataloging-in-Publication Data

Built-ins and storage / from the editors of Fine Homebuilding.
    p. cm. -- (Taunton's for pros by pros)
ISBN 1-56158-700-1
1. Built-in furniture. 2. Cabinetwork. 3. Storage in the home. I. Fine Homebuilding.
II. For pros, by pros. III. Title. IV. Series.

TT197.5.B8B82 2004
684.1'6--dc22

2004018184

Printed in the United States of America
10 9 8 7 6 5 4 3 2 1

The following manufacturers/names appearing in *Built-Ins and Storage* are trademarks:

Accuride®, Ben & Jerry's®, Blum®, Bosch®, Cheerios®, Enduro Poly™, Frisbee®, JDR Microdevices®, KV® 1385, Minwax®, NuTone®, Panasonic®, Plexiglas®, Polyshades®, Rev-a-Shelf®, Sub-Zero®, TiteBond®, Ultralume™, Waterlox®, Woodworker's Supply®

Special thanks to the authors, editors, art directors,

copy editors, and other staff members of *Fine Homebuilding*

who contributed to the development of the articles in this book.

CONTENTS

PART 3: KITCHEN, PANTRY & LAUNDRY ROOM STORAGE

PART 4: STORAGE SOLUTIONS & DESIGN IDEAS

INTRODUCTION

Real-estate agents know what pushes the buttons of most home buyers, and they know how to push them in the right order. In just about every house I've ever looked at as a potential buyer, the agent has opened the front door and remarked, "Wow, look at those beautiful oak floors," followed by, "and you won't believe how much storage space there is in this house."

That says a lot about our quest for order. Most of us want a place for everything, and everything in its place. But just where are those "places"?

Every house, big or small, has unused nooks and crannies just waiting to work a little harder for you. Drawers in toe kicks under the cabinets, triangular lockers under a set of stairs, or spice racks recessed into the space between wall studs are just a few examples of putting neglected spaces to better use.

This book is a collection of articles from *Fine Homebuilding* that deal with creative storage solutions. Our authors, who are all professionals in their fields, take you along in their search for space in a range of real-world projects.

No more unused nooks and crannies to colonize? "Building a Hutch" will give you a chance to build something from scratch and tailor it to the details that look best in your house. "Building a Window Seat," on the other hand, delivers a triumvirate of closet space, bench storage, and a serene place to curl up and read a book.

The ideas in this collection just might be enough to get that real estate agent to start the tour by saying, "You won't believe how much storage space there is in this house," and get to the oak floors later.

— Charles Miller,
special issues editor, *Fine Homebuilding*

Outfitting a Clothes Closet

■ BY GARY KATZ

Plain and simple. A single shelf with a clothes-hanging pole remains the most common closet storage arrangement today, especially in new construction.

A rosy resting place for the closet pole. Small plastic cups called rosettes screw into the side-wall cleats to hold the closet pole.

Midspan support. If the closet pole has to span more than 44 in., a bracket that supports the shelf as well as the pole is installed in the middle.

Clothes closets used to be really simple. A single shelf with a pole high enough to keep long dresses off the floor seemed to be all anyone ever needed. Maybe people didn't have as many clothes then, or maybe space wasn't at such a premium. These days, though, people are demanding much more from their closets.

That Good Old Shelf and Pole Might Be All You Need

The most common closet-shelf arrangement in American households is probably still a single shelf and pole (photo, left). With this system, the shelf is installed at 68 in. with the pole at about 66 in. from the floor, which is high enough to keep even the longest hanging clothes, such as coats and dresses, off the carpet.

I make most single-shelf arrangements out of medium-density fiberboard (MDF), and the layout is simple. First, I mark the shelf height of 68 in. and draw a level line on the back wall of the closet. Then I measure and cut the cleat for the back wall. The side-wall cleats have to hold the closet pole, so I cut them long enough to catch a stud

Double pole for shirts, jackets, and pants

40 in. to 42 in.

Minimum distance between shelf and pole: 1¼ in.

Single pole for long items

Maximum unsupported width of pole: 44 in.

Sweater shelves: 12 in. high.

Shelf sections should be 30 in. to 44 in. wide.

42 in.

66 in.

Shoe shelves: 7 in. high.

Bottom shelf: 16 in. off floor for boot storage.

Hanging clothes need 24 in. to any adjacent walls or shelving.

Sizing up clothes storage. A well-organized closet makes the most out of available space with single-pole as well as double-pole hanging areas and different-size shelves for a variety of purposes.

Try to standardize the sizes of the various sections for all the closets throughout the home. This will allow you to set repetitive stops on a chopsaw and to cut shelving kits for several closets at once.

for solid support. I make the cleats out of 1×4 so that they are wide enough to attach the rosettes that I install to hold the closet pole (bottom left photo, p. 4).

Back in the closet, I hold the back-wall cleat to the level line, and I attach the cleat with two 8d finish nails at each stud. Next, I level the side cleats with a torpedo level and then shoot them into place. The closet shelf is now ready to be dropped into place on top of the cleats.

For any span greater than 44 in., I install a shelf-and-pole bracket midspan, using a 10-in. piece of cleat stock as a backer block to hold the bracket away from the wall (bottom right photo, p. 4). If there is no stud for securing the backer block, I use screws driven into drywall fasteners that will be hidden by the bracket. The bracket is then screwed to the cleat, to the backer block, and to the shelf.

On the side cleats, I center the pole rosettes the same distance from the back cleat as the hook on the support bracket, usually 11½ in. to 12 in., depending on the bracket. Next, I measure and cut a wooden dowel for the closet pole and drop it into place. A screw driven into the pole through the support bracket completes the installation. I always predrill so that the screw doesn't split the pole.

Planning Lets You Maximize Storage and Production

These days, most homeowners want to make the most of clothes storage in their closets, so the old shelf-and-pole system won't do. Besides the single shelf and pole, which is needed for long, hanging items such as dresses and coats, other closet-storage arrangements include a double pole for shorter hanging items, such as pants and shirts, and banks of shelves for shoes and sweaters.

I try to incorporate a section of each type of storage in every closet. In general, I keep each section to between 30 in. and 44 in. Anything shorter is too narrow to be of much use, and anything longer requires additional support.

One of the first things I look at is the type of closet being outfitted. With a reach-in closet (usually 22 in. to 36 in. deep), the storage sections all have to fit on the back wall of the closet. In these cases, I try to divide the closet into three equal spaces. For example, an 8-ft.-wide closet can be split into three 32-in. sections. If a reach-in closet is less than 7 ft. wide, I often leave out one section, depending on the client's needs.

Walk-in closets, on the other hand, are bigger and deeper and have storage on more than one wall. Shelving options in these closets are less restricted. Walk-in closets come in many shapes and sizes, from large rooms that double as dressing rooms to deep, narrow spaces with L-shaped shelving. Most walk-in closets I work on are 5 ft. to 6 ft. deep. If a closet is to have storage on both side walls, it should be at least 6 ft. wide.

The homeowner's needs also play a part in determining the width of each storage section. For example, if a client has only a few long dresses or coats, I may opt to keep the single pole smaller to maximize the space for other sections.

Finally, I keep layouts as simple as possible, and I try to standardize the sizes of the various sections for all the closets throughout the home. Keeping section sizes the same from closet to closet allows me to set repetitive stops on my chopsaw and to cut shelving kits (dividers, cleats, shelves, and poles) for several closets at once.

MDF Shelves Are Not Adjustable

For new construction or for unfinished closet walls, I usually go with MDF shelving, and the first step is always careful layout). (MDF shelving and closet walls are usually painted at the same time, covering any pencil marks I might leave.)

I begin by drawing a plumb line for each vertical divider that separates areas of shelving and hanging clothes. I put an X on the side of the line where the ¾-in.-thick divider will land. Then I draw level lines for each shelf (photo, below). The disadvantage to this system is that the cleats are permanently installed, so spaces between shelves must be predetermined.

MDF shelving has to be laid out on the wall. Because MDF shelving is supported on permanently mounted cleats, the entire layout is done on the unfinished closet walls.

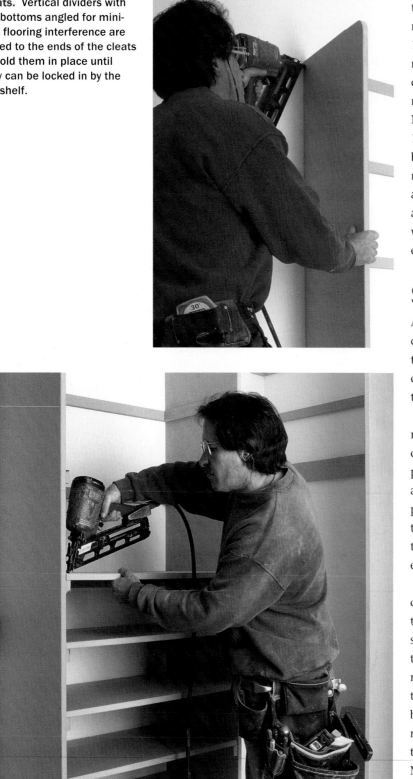

Dividers are nailed to the cleats. Vertical dividers with the bottoms angled for minimal flooring interference are nailed to the ends of the cleats to hold them in place until they can be locked in by the top shelf.

Shelves stabilize the dividers. Each shelf is nailed to the cleats permanently to help hold the dividers straight and square to the wall.

I cut the cleats, dividers, and shelves for the entire closet. Again, any cleat that carries a closet pole is made of 1×4, but I use 1×2 for cleats that support only shelves. I make my dividers 14 in. to 16 in. wide and cut the top corners at a slight radius for a nicer look. I usually cut the bottom of the MDF dividers at a 45° angle, leaving only 1 in. of the divider on the floor. This 1-in. bearing surface gives ample support without making a hassle for the flooring contractor, and it gives me a place to end the baseboard at each divider. A laminate trimmer fitted with a ⅛-in. round-over bit eases the sharp edges on the dividers.

Start in a Corner

After cutting, I stack all the cleats and dividers in the closet under the spot where they will be installed. Then I begin in one corner, installing the cleats on the lines that I drew.

When the first set of cleats is secured, I nail the first divider into the ends of the cleat and install the next set of cleats. This process is continued until all the dividers and wall cleats have been installed (top photo, left). Next, I glue and nail cleats onto the dividers for each shelf. I cut these cleats so that they end about ½ in. short of the front edge of the divider for a neat, clean look.

Once all the cleats and dividers are in, I cut and install the shelving. At this point, the dividers are still fairly flimsy, but the shelves help to hold them in place and keep them square to the wall. I install the shelves nearest to the floor first, nailing through the ends and the back edge into the cleats below. I then work my way up each section, nailing the shelves to the cleats as I go (bottom photo, left). Because the spacing of MDF shelves is determined ahead of time and because the shelves get caulked in and painted after the installation, I nail every shelf in place permanently.

Wooden biscuit connects the corner. A dry-fit wooden biscuit is inserted to keep the top shelf at the same level where two legs join in a corner.

Ready for the painter. With the top shelf nailed in to lock all the dividers in place, the last items to be installed are the rosettes and closet poles. The shelves will be painted at the same time as the closet walls.

I leave the top shelf for last. In most closets, the top shelf will turn a corner. You can buy aluminum H-clips that join the shelves at the corner, but these clips are unsightly even after everything on the interior of the closet is painted. Instead, I use a dry-fit wooden biscuit to join the inside corner (inset photo, above). With the shelf in place, I shoot a nail every foot or so along the back edge as well as a single nail into each divider along the front edge.

The last things to go in are the closet poles. I locate and install the rosettes and then cut each pole long enough so that it fits snugly into the rosettes, but not enough to push the dividers out of line (photo, above).

Wire Shelving: A Do-It-Yourselfer Alternative

One type of closet shelving gaining popularity today is wire shelving, especially among homeowners who like to do their own work. This product is prefinished and simple to install with few tools. Also, the open nature of wire shelving lets air circulate around clothes. Many building-supply outlets sell wire-shelving kits for standard-size closets, and many of these kits don't require cutting.

Admittedly, I haven't installed a whole lot of wire shelving, but when I have, I found that the instructions were always vague and misleading, so here's how I approach a wire-shelving installation.

In place of cleats, wire shelving is attached to the wall with plastic clips. Instead of solid dividers, the front edge of wire shelving is supported by poles also equipped with plastic clips. After screwing clips to the pole at whatever height I want the shelves, I use the poles to transfer shelf heights to the wall (photo, below left).

I then rest the back edge of the shelf against the wall and mark between the wires where I want the clips to fall. I put clips every 8 in. to 10 in., enough to support most clothing. With a level at the height mark, I now mark the exact location of all the clips. The clips are screwed into the framing wherever possible, and the rest go in with drywall anchors.

The shelving is prepared by first cutting it to length (if necessary) with either a hacksaw or with bolt cutters. Tiny flexible plastic caps slip over all the cut wire ends. The back edges of the shelves can now be snapped onto the clips, leaving the shelves hanging in place (photo, below right).

Wherever a shelf meets an adjacent wall, I install a special bracket that holds the front edge of the shelf. With the magnet of my torpedo level stuck to the shelf, I rotate the shelf into place and install the bracket. Next, I snap the shelves into the pole clips (photo left, facing page) and work on the shelving for any adjacent walls. Angled support brackets hold shelves level where they meet a shelf on an adjacent wall (photo right, facing page). Closet poles can be hung from any shelf with factory-made pole brackets, but special shelving with heavy wire on the outer edge is designed specifically for hanging clothes.

Poles take the place of dividers. Metal poles with plastic clips support the front edge of wire shelving. Here, shelf heights are taken from the pole.

Clips on the poles hold the fronts of the shelves. While clipped to the wall, the shelves pivot up and snap to the poles.

Angle braces support the shelves. Braces that snap into the wire shelving and screw to the wall reinforce long spans and support the shelves' ends where they meet in a corner.

Shelves snap onto the wall first. After clips are screwed to the wall, the shelving is snapped in and left hanging.

Prefinished dividers are set on a ledger. A 1x4 ledger, 16 in. from the floor, is attached to the closet wall to keep prefinished dividers off the floor.

The divider sets the height of the cleat. With a level ledger and dividers cut to the same length, the cleat for the top shelf is set flush to the top of the divider.

The only measurement in the closet. Other than the initial measurement for the ledger, the only measurement that needs to be taken is for the width of the remaining space. All other measurements are predetermined.

No Detailed Layout for Prefinished Shelving

When I'm asked to put a shelving system into an existing closet that has been painted, I opt for prefinished shelving, usually melamine. Heavy layout lines can't be drawn on the finished walls, so to avoid unsightly marks and blemishes, I let cleat and shelf sizes determine where the dividers are located. Also, because the shelving is prefinished, any cut edges that will be seen have to be edgebanded. With careful planning, I can minimize or eliminate edgebanding.

Most prefinished dividers have predrilled holes that accept shelf supports, eliminating the need for divider cleats. Unless the client requests dividers run all the way to the floor, I support the dividers on a 1×4 ledger along the wall 16 in. off the floor.

I cut all the dividers the same, and I cut extra dividers to hold shelves and poles on the side walls. I cut the shelves for all but the last section to the predetermined lengths. I also cut lengths of cleat for each section to hold the tops of the dividers and the top shelf.

I begin by setting the side-wall dividers on the ledger and nailing them to the wall (top left photo, facing page). Next, I nail the first top cleat to the wall. A divider is placed on the ledger and held against the cleat to set its height (top right photo, facing page). After the cleat is secured, the top-rear corner of the divider is nailed to the end of the cleat. The next top cleats and dividers are now nailed into place until I'm left with the last section. Up to this point, I have not made a single measurement or pencil mark in the closet except for the height and length of the ledger.

I measure and cut the top cleat and shelves for the last section and nail the cleat into place (bottom photo, facing page). The bottom shelf in each section squares and stabilizes the dividers, so it is attached permanently with special fasteners sold with the shelving material (bottom left photo). The back edge of the shelf is nailed to the ledger.

Near midheight of the divider, I set shelf-support brackets and lay in a single shelf. I predrill and drive a 1⅝-in. screw through the divider and into this shelf to stabilize the middle of the divider. Small decorative caps cover the screws (center photo). With the shelf holding the divider straight, I install corner brackets that secure the dividers to the back wall. Corner brackets also join the dividers to the top cleats.

After the dividers are secured and stabilized, I can space the rest of the shelves on the adjustable supports according to my client's wishes. Next come the poles. With prefinished shelving systems, I use brackets

Gadgets for prefinished shelving. Special pieces sold by shelving manufacturers help with installation. Shelf-attachment brackets (photo left) eliminate weak end-grain fastening. When screws are used, decorative end caps cover the heads (photo center). Oval-shaped steel closet poles hang on brackets that slip into predrilled holes (photo right).

that hold oval-shaped chrome-plated steel poles (bottom right photo, p. 13). The brackets pop into the predrilled holes, and the pole stock is cut to length with a hacksaw. If I don't think that the dividers have been stabilized adequately, I can run a cleat and shelf at every pole to keep the dividers from bulging and weakening. The final step is putting in the top shelf, which I nail or screw along the back edge as well as into the front corners of each divider (photo, below).

Gary M. Katz, author of The Taunton Press's The Doorhanger's Handbook (1998), is a carpenter, writer, and photographer in Tarzana, California.

Topping off a well-fitted closet. The top shelf goes in last and is attached to the cleat in the back and to the dividers in the front to tie the shelving system together.

Simple Closet Wardrobe

■ BY JIM TOLPIN

Closet storage systems are in vogue these days. Homeowners are buying prefabricated drawer units and paying closet specialists to install expensive epoxy-coated wire bins and shelves. Yet a handsome closet wardrobe can be constructed in just about a day by anyone who learns how to use a biscuit joiner and a shopmade layout jig.

A simple layout technique makes the process foolproof. The same approach is also effective for many other kinds of case pieces like bookshelves and cabinets. Plus, all of the work can be done on the job site with just a few tools and a workbench no more elaborate than a flat surface on a pair of sawhorses.

This surprisingly compact wardrobe unit (photo, right) features a bank of slide-out sweater drawers, adjustable shelves, and a hanging locker with a hat shelf. A typical bedroom closet is 2 ft. deep and 5 ft. or 6 ft. wide, so I make my wardrobes 16 in. deep and devote half the width of the closet to the drawers and the shelves. The size of the accompanying hat shelf and closet pole can be adjusted to fill the rest of the space.

Closet wardrobe made easy. A biscuit joiner and a layout jig make on-site construction of this space-saving closet wardrobe fast and reliable, which is an inducement for customers and an incentive for finish carpenters.

A Site-Built Wardrobe

Adjustable shelves, drawers, and a hat shelf make the wardrobe versatile. The base and the hat shelf are made separately, with the hat shelf sized to the width of the closet. The mitered base frame should be nailed and glued together with glue blocks added for rigidity.

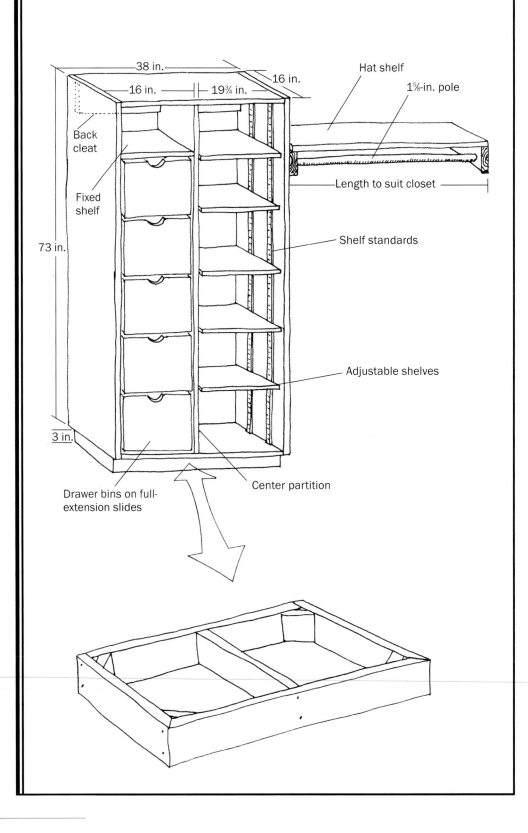

38 in.

16 in.

16 in.

19¾ in.

Hat shelf

1⅝-in. pole

Back cleat

Fixed shelf

Length to suit closet

Shelf standards

73 in.

Adjustable shelves

3 in.

Drawer bins on full-extension slides

Center partition

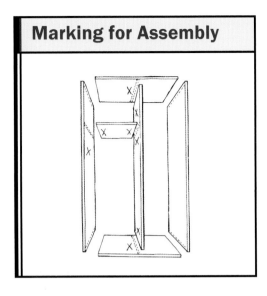
The project can be built from either ¾-in. boards or ¾-in. sheet stock like hardwood plywood or melamine (particleboard covered with a thin plastic laminate). Some suppliers stock melamine that is precut to 16-in. widths with a thin strip of the same laminate applied along one edge. This gives the piece a finished look.

Making the Base Frame

The wardrobe is essentially a tall box with a vertical divider down the center. One side is devoted to drawers and a fixed upper shelf, and the other to adjustable shelves. The case sits on a 3-in.-high base (drawings, facing page), with a 1½-in. toe space in the front and a 1½-in. overhang on each side. The base-frame height and the rear inset can be adjusted to clear any existing baseboard.

Start with the base, cutting the frame components to length and width for whatever size wardrobe you've decided to build. I rip sheet stock for the base, but you could also use 1×4s or even 2×4s.

Miter joints, which I use on the front corners of the base frame, may be made with an electric miter saw or a handsaw and a miter box. After the base pieces are cut to size, hold the front and back pieces together and mark on their top edges the location of the interior spacer at the middle of the base.

Finish nails and glue are adequate for assembling the cabinet base. You also may install glue blocks on the inside corners for additional strength. Biscuit joinery isn't necessary for the base because it is not subjected to much stress once the cabinet is installed, and using the joiner may actually take longer.

Cutting the Components

Cut the case sides, the top and bottom pieces, and the interior partition to length and width. Be sure the crosscuts are perfectly square, or the case may rack when assembled. I don't put a back on the case, so I use a cleat to help stiffen the case and to give me a way of attaching it to the wall of the closet. Cut a notch in the upper back corner of the center partition for a ¾-in. by 5½-in. cleat.

Shelf standards on the right-hand side of the case may be surface-mounted later. Or you can use a router and an edge guide at this point to cut shallow dadoes and let the shelf standards into the side of the case. But take care to stop the dadoes before they reach the top edge; otherwise, the ends of the dadoes will be visible after the case is assembled.

Lay out the location of the center partition on the top and bottom components. Also mark the location of the fixed shelf on the edge of the left side of the case and on the partition. Remember, however, the partition and the side pieces are unequal in height, which must be taken into account when marking the position of the fixed shelf. Also mark where top and bottom pieces will intersect the sides.

Before beginning the slotting process, dry-assemble the case on a flat surface. You may need to apply light pressure with pipe clamps to stabilize the structure. When all case components are aligned to their marks, check to be sure everything has been cut to the proper length and that the case is square. Lightly mark the case pieces so that you can assemble them quickly and correctly later (drawing, p. 17). I chalk an "X" on the faces of the components to help locate the slotting jig and to orient pieces to the bench top during the end-slotting process.

A Simple and Effective Jig

The case is held together by 32 biscuits, which means that 64 slots had to be cut, and they had to line up with each other. To cut the slots quickly and accurately, I built a jig (drawing, facing page) that locates the slots and helps orient and hold the biscuit joiner to the work. Positioned like a try square, the jig can be used from either the front edge or the back edge of the stock.

The entire jig can be made from ¾-in. hardwood plywood. It consists of an arm that is sandwiched between two crosspieces at a 90° angle. The crosspieces make up the base of the jig. The length of the arm extending beyond the edge of the base is the same as the width of the material you are joining—in this case 16 in. Once you have cut out the pieces for the jig, align them to a framing square and assemble them with glue and screws. But jigs can be made to any width and marked where you want to cut the slots. For the case of the wardrobe, I marked four evenly spaced slots for each joint.

Once the jig has been made, cut the slots in the side pieces of the case. These slots will receive the biscuits that are inserted into the ends of the top and bottom pieces. Position the jig back from the end of one of the case sides by the thickness of the stock and clamp the jig in place (left photo, facing page).

The centerline of the joiner's base should be aligned with one of the marks on the arm that indicates the center of a slot. Because the joiner is being used near the end of the workpiece, a scrap of case stock placed under the tool's face will add support and assure that the slot is perpendicular to the face of the work. This also is the time to slot the inside

faces of the top and bottom pieces to receive biscuits that are inserted in the ends of the partition. Position the jig to the X side of the mark you made to locate the partition on the top and bottom pieces and make the slots. Slot the fixed shelf the same way.

Biscuit joinery gives you some latitude in the placement of the slots. In side-to-side alignment, discrepancies of up to ¾ in. will not affect the outcome of the joint. But accurate vertical alignment is critical. Work away from the chip discharge of your biscuit joiner; otherwise, you will have to clear the work surface each time you reposition the tool for the next slot.

Cutting End Slots

Use the same jig to position the joiner for slotting into the ends of the top, the bottom, the partition, and the fixed shelf. For this operation, align the jig flush with the end of the piece to be slotted and clamp it securely in place (right photo, facing page). Orient the side of the piece marked with an X during dry assembly face down on the workbench and make sure the piece lies flat. The presence of chips or other foreign objects between the work and the bench, or warps in the bench itself, can easily misalign the end slots.

To slot the end of the stock, place the joiner flat on the work surface and orient the tool's centerline to one of the marks on the jig's cross arm. The base of the jig provides surface area for one of the joiner's antikickback pins, helping to hold the joiner in place. Again, work away from the expulsion of saw chips.

Assemble the Case

The first stage in assembling the case is joining the center partition to the top and bottom pieces. Apply glue to the slots (I use yellow glue), preferably with a special applicator tip on your glue bottle to keep the process mess-free. Then insert the biscuits and pull this I-shaped assembly together with pipe clamps. Be sure the components stay perfectly square to one another. Don't move this assembly until the glue is completely dry (at least a couple of hours).

The second stage is joining the rest of the components to the first-stage assembly. Assemble the case on a flat surface that can support the entire structure. Orient the marked front edges upward, making sure all the components are in their proper positions. Then apply the glue, insert the biscuits, and clamp the components together. Be sure the front edge of the boards are flush, and take diagonal measurements to the outside corners of the case to check for square.

Biscuit Joiner Jig

This shopmade jig makes reliable biscuit joinery a speedy process. It can be used from either the right or the left and includes reference marks that line up with the centerline on the biscuit joiner.

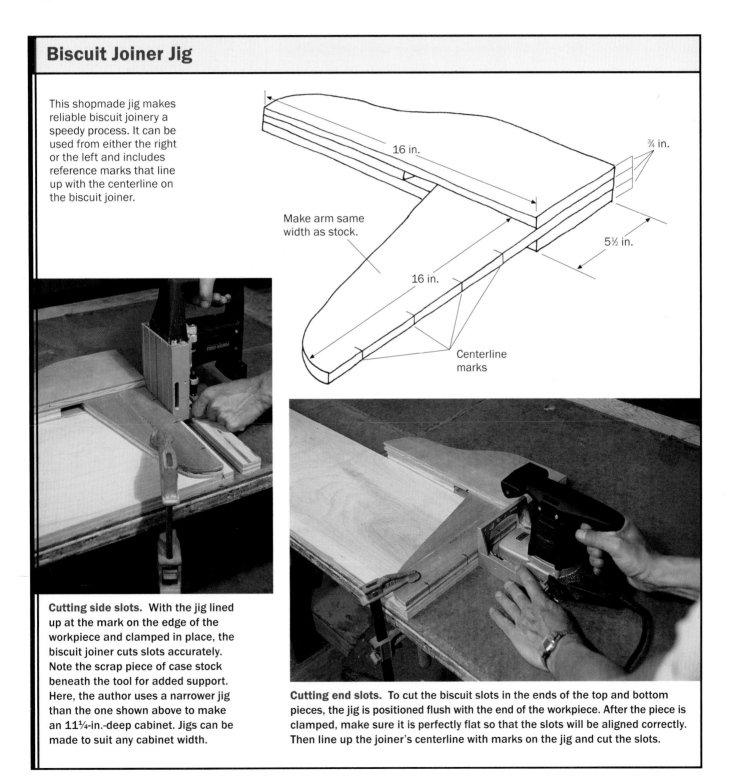

16 in.

¾ in.

Make arm same width as stock.

5½ in.

16 in.

Centerline marks

Cutting side slots. With the jig lined up at the mark on the edge of the workpiece and clamped in place, the biscuit joiner cuts slots accurately. Note the scrap piece of case stock beneath the tool for added support. Here, the author uses a narrower jig than the one shown above to make an 11¼-in.-deep cabinet. Jigs can be made to suit any cabinet width.

Cutting end slots. To cut the biscuit slots in the ends of the top and bottom pieces, the jig is positioned flush with the end of the workpiece. After the piece is clamped, make sure it is perfectly flat so that the slots will be aligned correctly. Then line up the joiner's centerline with marks on the jig and cut the slots.

Instead of clamps, 1⅝-in. drywall screws may be used to hold the case together while the glue sets. Screw holes made through the sides of the case should be counterbored and then plugged. Or screws can be capped with plastic covers. These plastic covers are available through most mail-order hardware suppliers.

When the glue has dried, remove the clamps. Flip the case over on its face edge, then glue and screw the cleat to the notch in the partition. The ends of the cleat will butt against the inside faces of the case, where they are glued and nailed using finish nails. For additional strength, install several screws through the top into the cleat, counterboring and plugging the holes or capping the screws.

Making the Drawers

The drawer components are cut, marked, and joined just like the case was assembled (drawing, facing page). Most drawer hardware requires a ½-in. space between the side of the drawer and the inside face of the case. Therefore, if the cabinet opening is 16 in., the drawers should be 15 in. wide. But check the specifications of the hardware you buy to see how much clearance the slides require. The drawer fronts are cut nearly as wide as the opening, which keeps the slide hardware out of sight when the drawers are closed. I use full-extension slides because they allow easy access to the entire depth of the drawer.

Cut the drawer parts to size (wait until later to cut the faces). Then dry-assemble the drawers and mark them for location and orientation. You can use the same positioning jig to cut the slots for the drawers, but you will have to make new slot-reference marks because the drawer pieces are not as wide as the case. Cover the arm of the jig with masking tape, then mark the locations for three biscuit slots on the tape. Use the jig to make slots in the face and the ends of the pieces.

Cut a ¼-in. by ¼-in. groove along the bottom inside face of the parts to receive the ¼-in. plywood bottoms. Then glue and insert the biscuits, run some glue inside the

groove, and assemble the sides around the bottom panel. Drawer bottoms that fit snugly will help keep the drawer square when it is clamped. Apply band or pipe clamps, wipe off excess glue, and check each drawer for square with diagonal measurements before setting it aside to dry.

Installing the Hardware

Next, put the case on one side and lay out and install the shelf standards and the drawer slides. Flip the case over and install the hardware on the opposite side. Secure the drawer hardware to the case sides using only the slotted holes; this will allow final adjustment later. A tick stick, a length of scrap lumber marked to show the location of each drawer slide, will help you lay out both sides of the drawer bay quickly and evenly.

Now it is time to install slides on the drawers. Cut the drawer faces and cut the semicircular hand grips in the top edges. Screw the faces to the drawers from the inside. If you use plywood for the drawer fronts, you will have to band the edges with a thin strip of solid material or veneer. Test the drawers by sliding them into the case on their slides. Adjust the gaps between the drawer faces by moving the slide hardware on the case sides up or down. When you are satisfied that the spaces between the drawers are even, secure the slides by adding screws to the round holes. Then cut the adjustable shelves to size and try them on the standards.

Installing the Wardrobe

Set the base frame assembly, still unattached to the case, into position on the floor of the closet. Carefully level the frame from side to side and from front to back. Shim where necessary between the bottom of the frame and the floor. Screw the leveled frame to the floor by running 2-in. drywall screws from inside the frame down at an angle. Locate

Making the Drawers

Drawers are made with the biscuit joiner and the jig shown on page 19, with ¼-in. plywood bottoms and applied drawer faces of solid lumber. Make drawers 1 in. narrower than the opening to leave room for the drawer slides.

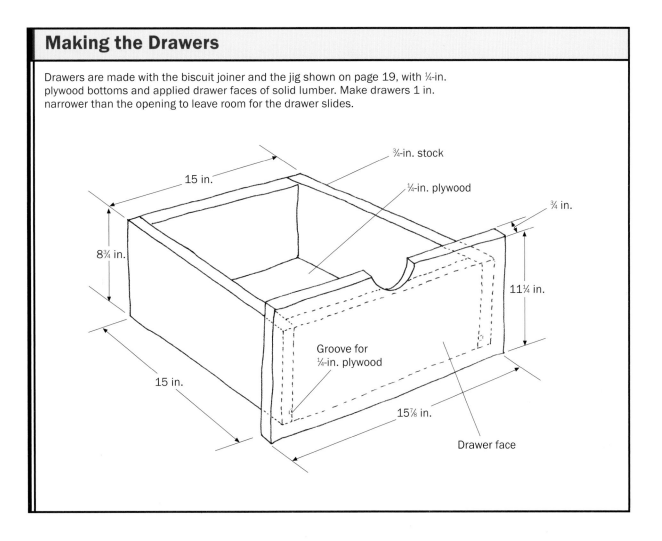

¾-in. stock

¼-in. plywood

¾ in.

15 in.

8¾ in.

11¼ in.

Groove for
¼-in. plywood

15 in.

15⅞ in.

Drawer face

the screws so that they will pass through the shims, but be careful that screws don't break through the face of the base. Score and snap off the shims (or saw them off with a veneer saw) where they protrude from the frame.

Remove the drawers and the adjustable shelves from the case to lighten the load, then set the unit on its base frame. Be sure the overhang of the case is equal side to side and parallel to the face of the base frame. Then join the base and the case together. Finish nails will hold the case permanently in place.

Fasten the wardrobe to the wall by running 2½-in. drywall screws through the back cleat into wall studs. But make sure the face of the wardrobe is square before securing it to the wall. Because the case doesn't have a full back, it could rack during installation if you're not careful. If the wall is leaning back out of plumb, slide shims between the cleat and the wall surface first.

The hat shelf, with closet pole below, is built to suit the closet width and is best tended to after the rest of the wardrobe is in place. Cut the two shelf cleats, then drill holes in them to accept the ends of the 1⅜-in. closet pole. In locating the holes on the cleats, make sure a hanger will clear the shelf and the back wall. Or you can use surface-mounted rosettes for the pole, which are commonly available from building-supply stores.

Simply screw one cleat to the wall and another to the side of the cabinet with the pole in between (cut the pole about 3/16 in. short to make its installation easier). Then cut the shelf to fit the opening, scribe the back and the far edge to the wall, and drop it into place.

Jim Tolpin is a woodworker and writer living in Port Townsend, Washington. He is the author of The New Cottage Home *and* The New Family Home, *both published by The Taunton Press, Inc.*

Custom Closet Wardrobe

■ BY PHILIP S. SOLLMAN

For the last three years my wife, Jeanne, and I tramped from our bedroom to the other side of the house to get to a closet so that we could get dressed. Probably because we built the house ourselves, we hadn't gotten around to the built-in clothes storage for the master bedroom.

Our bedroom is a 16-ft. by 26-ft. rectangle with the bed roughly in the middle, facing tall, second-floor windows that overlook a distant coppice to the south. The blank wall on the opposite end of the room was the obvious location for the closet wardrobe.

Over morning coffee in bed, Jeanne and I would discuss alternative designs for the much-needed closet wardrobe. Initially, I had envisioned an open dressing unit, which would be centered on the wall with enclosed closets on both sides. Jeanne balked at this idea, and she urged that everything be placed behind doors. I'd been hinting to Jeanne about deserving a new fly rod at the end of this project, so my strategy was not to make any waves. I didn't see the sense

Behind closed doors. The center unit opens to reveal banks of drawers below a marble counter and smaller drawers, sweater bins, and a mirror above. The lighted valance continues on the inside of the center doors.

Veneered plywood. Book-matched cherry veneers cover the plywood doors across the front of the wardrobe.

The integral pulls are routed into the doors along the wane edge of the veneer.

in opening a door to pull open a drawer, but I conceded that it would provide for a cleaner look and would allow me to do something interesting with the doors. All the while I was thinking about the fly rod this concession could produce.

Designing to Fit the Space

The wall to the left of the wardrobe has a large window that's about 4 ft. from the corner. I had crowned this window with a valance that concealed a curtain track. I stopped the valance 2½ ft. short of the corner in anticipation of the future wardrobe.

I decided it would look good to incorporate the valance into the design of the wardrobe (photo, pp. 22–23). This provided some visual continuity in the room and defined the height of the unit. The generous space left between the wardrobe and the ceiling would be perfect for displaying some of Jeanne's clay sculptures. She liked the idea, and I silently congratulated myself—I could feel that fly rod in my hand.

The exposed rafters in the cathedral ceiling are spaced 3 ft. 9 in. apart. Because of

their visual proximity to the wardrobe, this dimension determined the width of the individual compartments. There are four rafters, which meant that I ended up with five compartments. The three in the middle are of equal width; the two on the ends are about 20 in. wide.

A broad expanse of doors had the potential of looking plain and uninteresting, so I extended the center unit of the wardrobe past the flanking units. This created interesting niches to the left and the right, which could be lit from above by spotlights recessed in the valance. Extending the center unit also provided a terminus for the valances and increased the depth of the drawers in that unit. The center unit became the focal point of the wardrobe.

In the center unit I wanted a generous counter space with two banks of drawers below it and a mirror, small drawers, and shelves above (photo, pp. 22–23). I departed from rigid, right-angle geometry and introduced an inwardly curving front to the drawers and the compartments above. This reveals an inviting and elegant space when the doors are open. To enhance this quality, I continued the valance on the inside of the

Plywood Carcases Make Up Closet

Despite the sophisticated look of the finished wardrobe, the basic structure is simply five boxes made of ¾-in. cherry plywood.

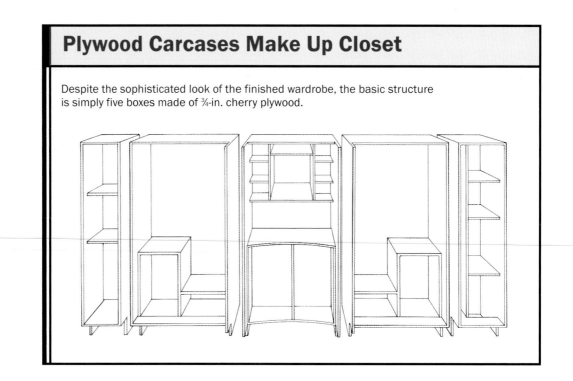

doors and wired them so that when they opened, two lights would turn on, automatically illuminating full-length mirrors set into the frame on the inside of each door.

Simple Carcase Construction

The 14-ft.-long wardrobe was too large to build as a single unit, so I built five individual sections (drawing, facing page) and trimmed them out when they were all assembled side by side in the bedroom. I allowed a ¾-in. space between sections, which eased installation and allowed space to run wiring to the lights and the switches.

Despite the sophisticated look of the finished wardrobe, the basic structure is quite simple. Using ¾-in. cherry plywood, I assembled five boxes with biscuit joinery and drywall screws. I used ¾-in. plywood for the backs, too, because it increased rigidity and gave me solid bearing for shelves and partitions. This part of the project was straightforward enough, but building the curved-front drawers for the center unit offered a bit more of a challenge.

Making the Curved-Front Drawers

I built the drawers of solid cherry but used ⁵⁄₁₆-in. by 3-in. T&G cedar closet lining for the drawer bottoms. I set the cedar into grooves cut into the drawer sides, fronts, and backs. I used cedar because of its nice aroma. But I was disappointed in the poor quality of the cedar closet lining. Much of it was too short or bowed, and I had to recut the badly fitting T&G joints. Unfortunately, this material comes prebundled in kraft paper, so you can't see what you're buying.

The eight square drawers in the flanking units of the wardrobe were put together using dovetails at all four corners, but the curved-front drawers in the center unit complicated things. Their drawer backs and sides

Routing recessed drawer pulls. The trick to routing recessed pulls in a drawer front that curves is to rout the pulls before cutting the curve, while the drawer front is still flat. Here a plywood template, clamped to the drawer, guides a router fitted with a straight-cutting bit.

are square to each other, so I joined them with machine-cut dovetails. But because the curving fronts form odd angles with the drawer sides, I decided to butt these pieces and join them with biscuits (football-shaped splines). I was careful to cut adjoining drawer fronts from the same piece of wood so that the grain pattern continues across both banks of drawers from left to right.

I cut the curved drawer fronts from 1-in.-thick stock—thicker than the sides and the backs—and I also glued a curved piece, shaped on the bandsaw, to the inside of the drawer fronts. This curved piece on the inside compensated for the stock I would be removing from the outside when I shaped the curve. By doing this I would still have enough stock from which to rout recessed pulls.

I routed the recessed pulls into the fronts before I shaped the curve so that I still had a flat surface to clamp my routing template against (photo, above). I did this process in two steps. First, I used a straight-cutting bit to rout the recess, then I switched to a finger-pull bit, running it only along the bottom to undercut the edge and provide a grip.

Vacuum-Bag Veneering

Veneering involves bonding a thin (usually decorative wood) material to a rigid and stable substrate (like plywood). I used to do all of my veneering using the popular neolithic method of piling concrete blocks on top of my work and clamping the edges with every clamp I had available. Needless to say, this is inefficient, slow, and usually leaves areas that didn't bond well, particularly in the middle of a broad surface where clamps aren't able to reach.

When I learned a couple of years ago that vacuum-bag veneering was not only available but also affordable to a one-man shop such as mine, I decided to throw my concrete blocks back on the brick pile. This veneering method proved to be simple, quick, effective, and very user-friendly.

Vacuum-bag veneering involves two basic components: a small vacuum pump and a vinyl bag that contains the pieces to be bonded together. The pump evacuates the air from the bag, which allows the atmospheric pressure to press the veneer and its substrate together.

The pump has a gauge on it that measures the atmospheric pressure inside the bag. At 21 in. to 25 in. of mercury (Hg), the force is 1,764 lb. per sq. ft., and the pump shuts off automatically. If the pressure drops, it cycles back on. This pressure is applied evenly across all surfaces—top, bottom, and sides.

I bought my vacuum-bag setup two years ago for $835. It came with a large 30-mil vinyl bag opened at both ends. These ends can be sealed using plastic tubes

A vacuum-bag veneer press consists of a vinyl bag and a pump that evacuates air from the bag. Pieces to be veneered, like the plywood doors and the cherry veneer above, are covered with glue and put in the bag. The pump exerts a pressure of more than 1,700 lb. per sq. ft.

and C-shaped channels, which are provided. The top of the bag is clear so that you can see the work being pressed. The bag has a hole at the bottom of one end with a plastic sleeve that connects to the pump.

My bag accommodates a 4-ft. by 10-ft. platen (a sheet of ¾-in. plywood covered with plastic laminate), which is grooved to facilitate the evacuation of air from the bag. The system won't work without this platen.

The manufacturer recommends that the veneer not touch the bag or the platen, so use a caul. This caul can be any sheet material at least ⅛ in. thick. Normally the veneer and its substrate are placed upside down on top of the caul. Because I cut my veneers thick (about ⅛ in.), however, I have good results without the caul, and I place my door panels in the bag right-side up.

I spread yellow glue evenly over the surface with a 6-in.-wide putty knife. I usually leave my work in the bag overnight. Large panels, such as the doors for my closet wardrobe, will warp when removed from the bag but will flatten out again after the glue is thoroughly dry. This warping is a result of the water-based glue.

It only takes a few minutes to draw a vacuum once the bag is sealed, so there is still time to reset the veneer if you don't line it up quite right. I often use two wire brads nailed through the veneer into the substrate, which keep the veneer from moving when it is placed in the bag (or you can use masking tape). These brads must have flat or rounded heads and must be driven nearly all the way into the surface so that they won't damage the vinyl.

Sources

Woodworker's Supply®, Inc.
5604 Alameda Pl., NE
Albuquerque, NM
87113
(800) 645-9292
www.woodworkers.com

Waterlox Coatings Corp.
9808 Meech Ave.
Cleveland, OH 44105
(800) 321-0377
www.waterlox.com

Vacuum Pressing Systems, Inc.
553 River Rd.
Brunswick, ME 04011
(207) 725-0935
www.vacupress.com

I shaped the faces of the drawers by clamping together each bank of assembled drawers, then power planing and belt sanding the final curve into them (top photo). This guaranteed that I would have a smooth curve across all the drawers.

Using Wooden Drawer Slides

I think wooden drawer slides look better than metal slides, and they work just as well. I made my drawer slides out of ¾-in. by ⅜-in. strips of cherry that are screwed to the plywood sides of the drawer compartments. The slides fit into slots routed into the sides of the drawers.

The key to making wooden slides work is aligning them so that the drawers don't bind. To do this, I use a simple template, or story pole. I made the story pole by first stacking all the drawers together with shims in between to account for clearances. Then I positioned a thin strip of wood (the story pole) against the drawer sides and marked the locations of the grooves with a utility knife on the story pole (left photo, facing page). Next, I took my story pole to the bandsaw and cut out small notches at my marks, which allowed me to rest the wooden slides in the notches as I screwed the slides to the plywood (right photo, facing page).

Making the Doors

With the exception of the two doors on the center unit, all the doors are constructed of ¾-in. cherry plywood. Because the center-unit doors would have mirrors bonded to their inside surfaces, I decided to make them from less-expensive birch plywood. Cherry plywood has incredibly thin and vulnerable

Shaping curved drawer fronts. The two banks of large drawers in the center unit feature a gentle curve across their fronts. To obtain a smooth curve, the author clamped the entire bank of drawers together, then power planed and sanded their curves simultaneously.

veneers, which made it unsuitable for use on the outside of the doors. I decided from the beginning to laminate my own, more substantial veneers to this substrate.

I resawed the 18-in.-wide cherry veneers on the bandsaw in my shop. By feeding the stock very slowly into the blade, a veneer of surprisingly consistent thickness could be produced. These rough veneers were sanded down to about ⅛ in.

The planks that produced these wide veneers came from logs that I milled and air-dried myself. During the milling of these logs, I had preserved the beautiful wane edge and intended to make use of it as part of a recessed door pull.

I edged the plywood panels with solid cherry all around and laminated my veneers to them using TiteBond® glue and a vacuum-bag veneer press (see sidebar, pp. 26–27). After the doors were veneered, I once again used

my finger-pull bit in my router and followed the natural contour of the wane edge.

The doors were sanded and hung. And the entire unit was finished with Waterlox® Transparent and Waterlox Satin Polyurethane.

Because I haven't gotten my fly rod yet, I asked Jeanne how I was going to finish this article. She sympathized and said I had a real problem. Oh well, maybe after I build the garage.

Philip S. Sollman is a woodworker in Bellefonte, Pennsylvania.

A story pole for drawer slides. The author stacks drawers with shims in between to account for clearances. Then he locates the drawer-slide grooves on a story pole (left). After cutting notches at each mark, he uses the pole to position the slides as he screws them into the cabinet side (above).

Building a Fold-Down Bed

■ BY PATRICK CAMUS

To transform the bed, the author pulls the roll-out footboard from the wall (above), unlocks the hinged platform from its place between the bookcases, and lowers it onto the footboard (left).

Source

H. B. Ives–Schlage
111 Congressional
Blvd., Ste. 200
Carmel, IN 46032
(800) 766-1966
www.schlage.com
Sliding bolts

Down for the night. Tucked away, the bed masquerades as a wall; the wainscot details align with the adjacent cabinet doors.

The applied molding at the bed's foot fits into notches in the support.

Lynnette and I don't have that many houseguests, but the few that we do have tend to stay for long periods of time. We also don't have that much room: Our townhouse is narrow, and space is always a big concern. Our only spare room is just big enough for a bed, but we thought it would be nice to use the room as a dressing room when it wasn't occupied by the odd cousin or in-law. A Murphy-style bed seemed like just the thing.

After some research, I quickly became disillusioned with everything premanufactured, which seemed bulky and lacking in finesse. I wanted something that could truly blend into and even enhance the room, something that off-the-shelf choices never do.

Because the room measures only 9 ft. by 9½ ft., it was important that the bed be as slim as possible when folded away. I also wanted the bed's underside to be flat and smooth so that it would look like the other three walls (drawing, p. 32). I found a lumberyard that stocked 5-ft. by 10-ft. sheets of medium-density fiberboard (MDF), which made a perfectly sized mattress platform. Flanking the bed with shelving units kept the massing down while forming a perfect niche for the bed. The bed could then be locked into an upright position using sliding bolts buried in the shelves.

The support for the foot of the bed was the final item to tie together the design. Fold-down legs wouldn't maintain the flat surface of the platform, so instead, we opted for a roll-out footboard. This roll-out footboard is actually a box, big enough to store a comforter; the two flanking cupboards below the bookcases hold the pillows.

Disguised as wainscoting when the bed is stored in the wall, the footboard conceals the ledger and hardware. Although the MDF makes the bed feel heavy when it's descending, we've had only compliments about the bed's comfort. And when the bed is closed up, we're left with extra floor space that we can always use in a house that's 10 ft. wide.

Patrick Camus, an architect in Alexandria, Virginia, also builds furniture part-time.

The Mechanics of a Disappearing Bed

This Murphy-style bed was added to an existing room and depends on the flanking bookshelves for locking support and camouflage. The tongue-and-groove detailing of the roll-out footboard mimics the lines of the adjacent cabinet doors.

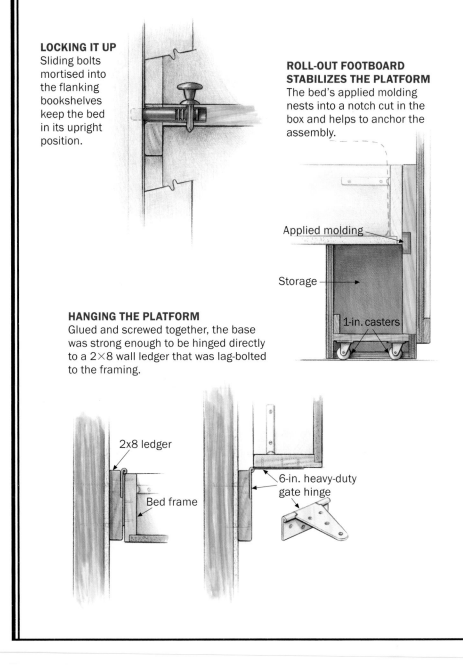

LOCKING IT UP
Sliding bolts mortised into the flanking bookshelves keep the bed in its upright position.

ROLL-OUT FOOTBOARD STABILIZES THE PLATFORM
The bed's applied molding nests into a notch cut in the box and helps to anchor the assembly.

Applied molding

Storage

1-in. casters

HANGING THE PLATFORM
Glued and screwed together, the base was strong enough to be hinged directly to a 2×8 wall ledger that was lag-bolted to the framing.

2x8 ledger

Bed frame

6-in. heavy-duty gate hinge

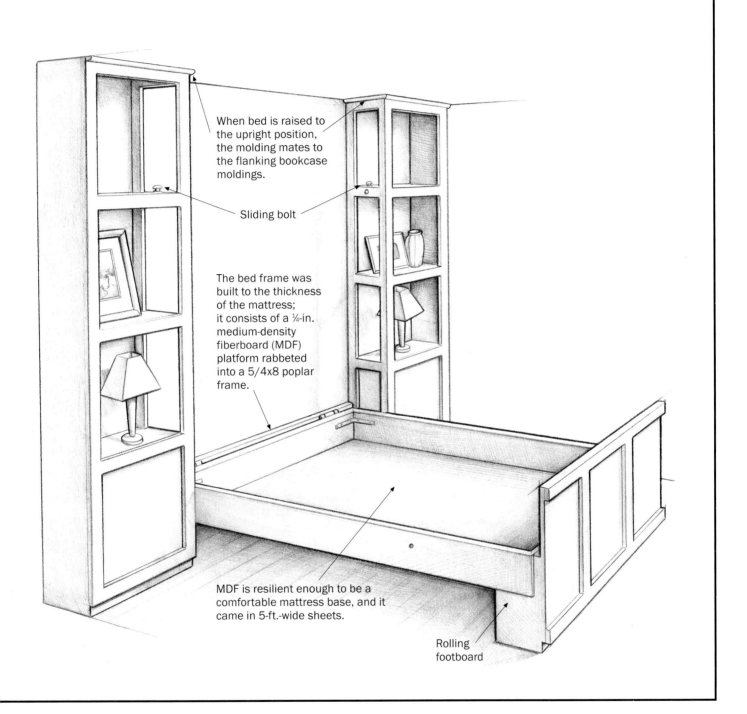

When bed is raised to the upright position, the molding mates to the flanking bookcase moldings.

Sliding bolt

The bed frame was built to the thickness of the mattress; it consists of a ¼-in. medium-density fiberboard (MDF) platform rabbeted into a 5/4x8 poplar frame.

MDF is resilient enough to be a comfortable mattress base, and it came in 5-ft.-wide sheets.

Rolling footboard

Fanciful Built-In Beds

■ BY JEAN STEINBRECHER

As I paged through Harriet DeWolfe's collection of books on Scandinavian houses, I marveled at the built-in beds. They are delicious hybrids—part furniture, part room, and part fantasy world. You don't just sit on one of these beds. They're too tall for that. Instead, you sort of occupy one, like a scout establishing an outpost on a piece of high ground.

Harriet and her husband, Russell, live in a house I designed for them on the western shore of Whidbey Island, in Washington state. From the outset, they were definite about the character they wanted in their home. It had to have Swedish flavor, color, and texture. And it had to have built-in cabinet beds, like the ones that Harriet and Russell have so enjoyed during their trips to Scandinavia.

I was fascinated by their vision of the house and excited by the prospect of doing some innovative design work based on historic precedent. It seemed a logical extension of my former career as a costume designer for the stage and my studies of historic preservation in architecture school.

The Master Suite Overlooks the Shipping Lanes

We eventually worked three built-in beds into the house. The first is in the master suite (photo, p. 39). This bed's alcove is raised high enough to allow a full view down the west side of the island, where ships follow the deep channel into Seattle. At the head of the bed, a shelf holds odds and ends and forms a backrest for in-bed reading. Along the sides, built-in night tables house drawers over cabinets. Built-in shelf units, replete with light fixtures and electrical outlets, flank the bed head and frame the window above. At the foot, bookshelves built into the lower platform are backed by deep storage drawers that open onto each side of the bed.

Like the trim in the rest of the room, the bed is painted off-white and ties into the chair rails and wainscoting. Quarter-circle brackets hold the foot of the mattress and echo the brackets holding the shelf at window-top level in the kitchen.

Inside the Cloud Bed

As a true built-in, this bed relies on the frame of the house for much of its own structural integrity. Cleats affixed to the wall hold up two sides of the mattress platform, and the house framing serves as anchorage for the arched sides of the bed. The box spring rests on a sheet of plywood supported by plywood bulkheads that house the three drawers under the bed.

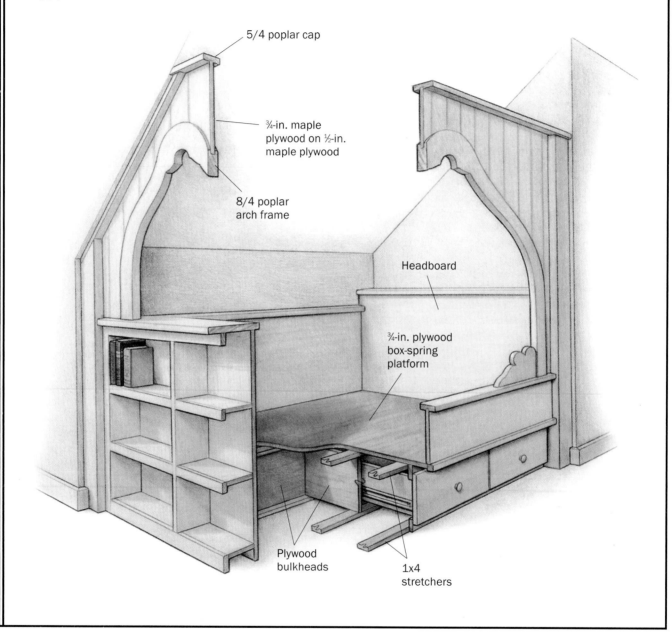

5/4 poplar cap

¾-in. maple plywood on ½-in. maple plywood

8/4 poplar arch frame

Headboard

¾-in. plywood box-spring platform

Plywood bulkheads

1x4 stretchers

Upstairs, the Beds Engage Sloped Ceilings

The DeWolfes' house has a cross-gable roof with shed dormers and cathedral ceilings. These quirky rooflines create grandmother's-attic-like spaces that are perfect for built-in beds.

Russell's study has a bed made of Douglas fir with turned columns and two-tone staining (photo, right). We based this bed on a historic example we found in *Scandinavian Country* by JoAnn Barwick (Clarkson Potter, 1991).

The bookcase at the tall end of this bed becomes the headboard. It is flanked by massive corner posts made of kiln-dried fir 6×6s. Our cabinetmaker, Dick Kieffer, had three posts turned at a local millwork shop. One post was split in half, becoming the two pilasters at the foot of the bed. Three deep drawers store linens and office supplies below the bed box made for a queen-size mattress and box spring.

Cloud Forms Meet Ogee Arches

Guests who stay in Harriet's study enter the queen-size sleeping nook, fondly called the cloud bed, through a sensuous ogee arch (photo, p. 35). The bed takes its name from the cloud-motif brackets that retain the mattress. We adapted it from a photo found in Elisabeth Holte's *Living in Norway* (Abbeville Press, 1993).

Dick made all the beds in his shop as individual components, then assembled them on site. For example, the cloud bed is made up of five parts: the drawers and their carcase; the headboard; the footboard and its bookcase; and the sides with the arched openings.

Bookcase as headboard. Vertical-grain Douglas fir, tinted with gray and vermilion, is the predominant material in the captain's bed (above). Cabinets and bookcases flanking the head of the bed in the master suite (photo, p. 39) provide storage for bedtime books.

Making the beds in sections made transporting the various parts into their appointed rooms easy—or at least possible. In fact, Dick met our client Russell for the first time on the stairs, where Dick had become wedged against the stairwell by the big-arch side of the cloud bed he was trying to maneuver around a corner. The two of them were able to wrestle the bedside back down the stairs. Fortunately, the windows weren't installed upstairs, so Dick and our builder, Phil Stringer, brought the bed components in through a second-story window opening.

The cloud bed is almost exclusively maple plywood and poplar trim (drawing, p. 36). To begin, Dick routed V-grooves 5 in. o.c. into sheets of ¾-in. maple plywood. Then he glued and stapled the grooved pieces to other sheets of ½-in. plywood running the opposite direction, making panels large enough for the bed front and side.

Working at full scale on paper, he sketched the arch and refined one side of the curved opening, then made a plywood template of the half-arch. Using the template as a guide, Dick glued up sections of 8/4 poplar into the rough shape of the wide trim pieces that define the arched openings. Each poplar arch blank is composed of five separate boards connected by scarf joints and using pairs of ½-in. dowels.

Dick used the template to mark cutlines on both sides of the arches and roughed them out on the bandsaw. Next, he cleaned up the edges of the arches using a router with a bearing-guided bit following the profile of the template.

The rest of the bed box is made of ½-in. plywood and trimmed with poplar. Dick sized the bed box to accommodate the DeWolfes' queen-size mattress and box-spring set, adding 1½ in. around the edges to allow space for bedding and bed-making. If we had used a futon or European mattress without a spring, we would have raised the bottom of the bed box and added ventilation holes in the top surface to allow the mattress to breathe.

The plywood drawers beneath the bed are 25 in. wide, 8 in. tall, and 28 in. deep. They slide on Accuride® 4034 full-extension 28-in. drawer glides with ball bearings and 1½ in. of overtravel. Dick swears by these drawer glides. They're rated at 150 lb., they're smooth and quiet, and he's never had a callback on them.

The drawer unit is set back 3¼ in. to create a kick space under the bed. This is the kind of detail that doesn't seem important until you've stubbed your slippered (or bare) toes on a pedestal bed without a kick space.

Once the components were in place, Dick screwed them together with 3-in. coarse-thread drywall screws. The wall cleats are affixed to the house framing, as are the plywood sides of the bed. They're screwed to the studs from the top, where the poplar caps conceal the fasteners. After plugging the screw holes and puttying any remaining divots, Dick spray-painted the bed in place with a coat of oil-based primer followed by two coats of satin oil-based paint. He kept the overspray to a minimum by tenting the bed and using a high-volume, low-pressure sprayer.

Guests Don't Want to Leave Now

The best feedback about these whimsical beds comes from overnight guests. Delighted to be hopping up onto beds that are a little too tall, protected from the big, bad world, guests of all ages become children again. They want cookies and milk and bedtime stories. They want to be tucked in and wished pleasant dreams. Best of all, they want to return for another visit.

Jean Steinbrecher, AIA, is based in Langley, Washington and specializes in residential design.

Bed Alcove

■ BY TONY SIMMONDS

When the middle one of my three daughters grew too old for the loft bed I built for her, the youngest, Genevieve, was happy to inherit it. The loft is in a small bedroom on the second floor of our house in Vancouver, B. C., Canada. Like many second floors of old houses, this one is really a half story, with sloped ceilings where the rafters cut across the intersection of wall and roof. The bedroom has only about 80 sq. ft., so its bed had to be on a raised platform to leave space for a dresser and a desk below.

Soon after she moved into the loft, however, Genevieve started bumping her head on the ceiling over the bed. When she eventually moved the mattress to the floor, I knew it was time for the old bed to go and for a new one to take its place. The bed alcove shown in the photo at right was the result.

Will It Fit?

The kneewalls that defined the sides of the room had originally been a little more than 6 ft. high, leaving a great deal of wasted space behind them. I proposed to recover this space by moving the kneewall over 4 ft. to accommodate a 3-ft.-wide mattress and a bedside shelf beyond that. Given the 12-in-12 pitch of the roof, this would bring the ceiling down below 3 ft. at the new kneewall. Would this be claustrophobic? To answer the question, I mocked up the space with packing crates and plywood to make sure there would be room to sit up in bed. A high ceiling is not a necessity over a bed—within reason, the reverse is true: A lower ceiling increases the sense of shelter and enhances the cavelike quality humans have always favored. Furthermore, a bed in an alcove that can be closed off from the rest of the room has qualities of privacy and quiet that are difficult to achieve in any other way. To get that extra layer of privacy, Genevieve and I decided that her bed alcove should have four sliding shoji screens.

The 9-ft. length of the space would provide room for a dresser and a vanity of some sort, as well as the bed. Drawers underneath the platform would triple the existing storage space. Light and ventilation would come from an operable skylight over the bed.

Tight fit. Into this 9-ft.-long space, the author squeezed a single bed, a row of 30-in.-deep drawers, a bookshelf and a vanity. A recessed fluorescent fixture illuminates the mirror from above while the vanity table is lit by a lamp behind the mirror. The baseboard reveals the line of the original wall. Above it, drawer fronts cut from a single 1×10 are screwed from behind to the drawers.

I had some misgivings about the location of this skylight in spite of the obvious benefits it would confer in terms of light and space. Having never slept directly under one myself, I didn't know whether a skylight so close to a bed would make sleep difficult. But in the end I was seduced by three arguments. First, the skylight would face north and therefore would not be subject to heat-gain problems; second, it would illuminate the shoji from behind; and third, there was the emotional pressure from my client—some drivel about the stars and the treetops and falling asleep to the sound of rain on the glass.

Tight Layout

Juggling existing conditions is the challenge of remodeling. None can be considered in isolation. For example, I had to decide whether or not to keep the existing 7-in.-high baseboard. I could have moved it, but I wanted to leave it in place, partly for continuity and partly to avoid as much refinishing as possible. Starting the drawers above the baseboard also meant that the baseboard heater already on the adjoining wall wouldn't have to be moved to provide clearance for the end drawer.

Four drawers fit into the space between the baseboard and the mattress platform. The drawers are 7 in. deep (6½ in. inside), which is ample for all but the bulkiest items. This brings the mattress platform to a height of about 18 in. With a 4-in.-thick mattress on top of it, the bed still ends up at a comfortable sitting height.

In plan, the mattress takes up almost exactly three-quarters of the 9-ft.-long space. The leftover corner accommodates a make-up table with mirror above and more drawers below. I imagined that the shojis would draw a discreet curtain over the wreckage of eyeliners, lipsticks, mousse, and everything else that was supposed to go in the drawers but never would.

I knew that this vanity area, and especially the mirror, would need to be lit, but beyond making sure there was a wire up there somewhere, I didn't work out the details during the preliminary planning. I was in my fast-track frame of mind at this stage of the project.

Site-Built Cabinet

The underframe of the bed is a large, deep drawer cabinet. You could have it built by a custom shop while you get on with framing, wiring, and drywalling. Custom cabinets are expensive, though, and after nearly 10 years in the business of building them, I appreciate the virtues of their old-fashioned predecessor, the model A, site-built version. It's economical in terms of material and expense, and you can usually get a closer fit to the available space.

The partitions supporting my daughter's bed are made from ⅜-in. plywood sheathing left over from a framing job (the rewards of parsimony). Each partition is made from three layers of sheathing (drawing, left). The center layer runs the full height of the partition, but the outer ones are cut in two, with the drawer guide sandwiched between the top and bottom pieces. The guide is simply a piece of smooth, fairly hard wood, ¾ in.

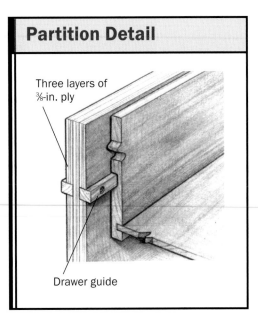

Partition Detail

Three layers of ⅜-in. ply

Drawer guide

Bed-Alcove Anatomy

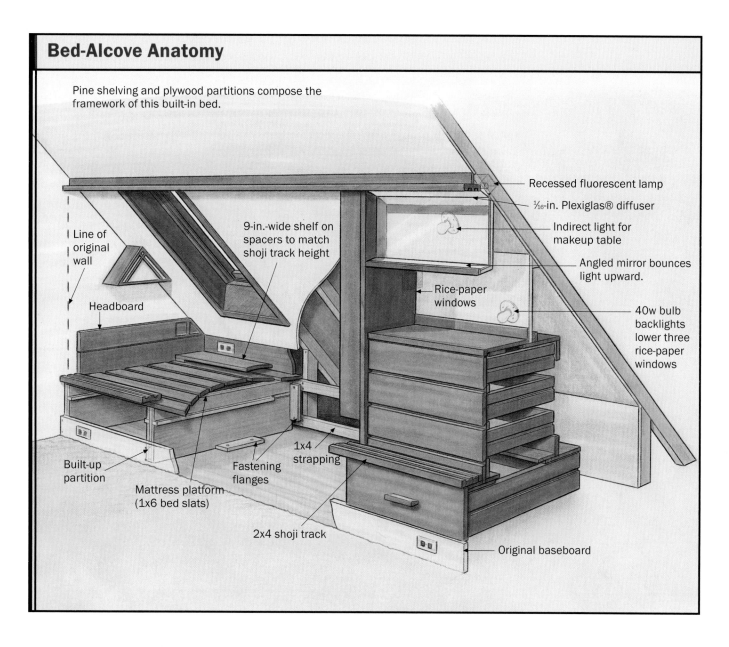

Pine shelving and plywood partitions compose the framework of this built-in bed.

Line of original wall

Headboard

9-in.-wide shelf on spacers to match shoji track height

Recessed fluorescent lamp

$\frac{1}{16}$-in. Plexiglas® diffuser

Indirect light for makeup table

Angled mirror bounces light upward.

Rice-paper windows

40w bulb backlights lower three rice-paper windows

Built-up partition

Fastening flanges

1x4 strapping

Mattress platform (1x6 bed slats)

2x4 shoji track

Original baseboard

thick and wide enough so that it projects $\frac{3}{8}$ in. into the drawer space.

Unless circumstances demand the use of mechanical drawer slides, I prefer to hang drawers on wooden guides. I have provoked derision from cabinetmakers because I use wooden guides in kitchens, but when it comes to bedrooms I am almost inflexible. Even large drawers like these will run smoothly year after year if they are properly fitted and if the guides are securely mounted. And for me there is a subtle but important difference between the sound and the feel of wood on wood vs. even the finest ball bearings.

I attach the guides with screws rather than with glue and nails so that they can be removed, planed, and even replaced without difficulty should the need arise. A groove in the partition to house them is not necessary, but it's a way of ensuring that they all end up straight and exactly where you want them.

For this job, the pairs of guides on the three middle partitions had to be screwed to one another, right through the core plywood. I drilled and counterbored all the screws and clamped the partition to my workbench to make sure everything stayed tight while I drove the screws. Then, with the partition still on the bench and after inspecting every screw head carefully for depth below the surface, I set the power plane for the lightest possible cut and made three passes over each guide: first over the back third only, then over the back two-thirds and, finally, over the whole length of the guide. Tapering the guides so that they are a fraction farther apart in the back allows the drawer to let go, rather than tighten up, as it slides home.

Partition Alignment

Installing the partitions is the trickiest part of a site-built cabinet job like this one. I said earlier that you could save on materials by building the cabinet in place, but you can't save on time. After all, anyone with a table saw can build a square cabinet in the shop, but building one accurately in a closet or in an unfinished space under the rafters takes patience and thoroughness. The key to success is to establish a datum line, then lay out everything from this line, leaving the wedges of leftover space around the perimeters to be shimmed, trimmed, fudged, and covered up as necessary.

In Genevieve's room, the existing baseboard provided a datum line in both horizontal and vertical planes. First, I divided the baseboard's length so that the four drawer fronts would lie directly below the shoji screens. I ran one screw into each supporting partition, about 1 in. below the top edge of the baseboard. Then I plumbed the front edge of the partition and secured it with a second screw near the bottom of the baseboard. With the front edges located and the partitions standing straight, the next job was to align them to create parallel, square openings.

I built the new kneewalls 48¾ in. back from the inside face of the baseboard. This allowed me to run a couple of 1×4 straps horizontally across the studs to provide anchoring surfaces for the 48-in. partitions (drawing, p. 43).

To align the partitions, I used hardboard cut to the full opening width (photo, below). As long as the hardboard is cut square, and the partitions are secured so that the hardboard fits snugly between them, the resulting opening will also be square. I used screws to fasten the plywood flanges that held my partitions in place, just in case adjustment should be necessary.

Aligning partitions. Load-bearing partitions made of three layers of ⅜-in. plywood separate the drawer bays under the bed and support the mattress platform. The drawer guides are sandwiched between the outer layers of plywood. The photo shows the hardboard panels that helped to align the partitions. Once the panels were in place, the partitions were screwed first to the baseboard and then to strapping along the stud wall. The 1×2s clamped to the leading edges of the panels are gauges that will be used to determine the depth of the grooves in the drawer sides.

Linked by slat. The partitions are tied to one another across their tops by 1×6 pine slats. Spaces between the slats provide ventilation for the mattress. At the right side, the carcase for the vanity drawers sits directly atop the bottom drawer partitions.

When all the partitions were in place, I cut pieces of 1×2 to the exact dimension between each pair of drawer guides. Centered on the drawer fronts, the 1×2s are gauges that show how deep the grooves need to be in the drawer sides.

The bed slats also act as ties to link all the partitions together (photo, above). I used dry 1×6 shelving pine for the slats, but almost anything that will span the distance between supports will do. I left an inch between the slats to keep the mattress well aired. I learned this the hard way when an early bed I built on a solid plywood platform developed mildew on the underside of the mattress cover.

Fitting the Drawers

Before putting anything on top of the platform, I built and fitted the drawers. The drawers have ⅛-in. clearance between their sides and the partitions. The ⅜-in. projection of the drawer guide thus creates a ¼-in. interlock with the sides. All the drawers are 30 in. deep, but I let the sides extend 6 in. beyond the back of the drawer. The extensions support a drawer right up to the point where its back comes into view.

If time and budget allow, I use a router jig to dovetail the front of a drawer to its sides, but the back just has tongues cut on each end that are glued and nailed into dadoes in the sides (I take care not to put any nails where the groove for the guides will be plowed out). The drawer bottom rides freely in a groove cut in the front and the sides and is nailed into the bottom edge of the back, which is only as wide as the inside height of the drawer. Fastening the bottom here helps to keep the drawer square.

Fitting the drawers should present few problems if they are built square and true and if time and care have been invested in positioning the partitions. Don't try for too tight a fit, especially in the width of the groove. My guides were ¼-in. material, and I plowed out a ¹³⁄₁₆-in. dado in the drawer side. They're not sloppy.

On the other hand, you should be more stingy about the depth of the grooves. Remember, the guides have been planed to allow increasing clearance as the drawer slides home. Too much slop here can cause the drawer to bang about from side to side and actually hang up on the diagonal. You can always plow a groove out a little deeper. A router with a fence or a guide attached is the ideal tool for this because you can easily make very small adjustments. If things go wrong, you can glue a length of wood veneer tape into the dado, but it's nicer not to have to do that.

I dress the groove with paraffin wax, but only when I'm sure the drawer doesn't bind. Patience in working toward a fit has its reward here. The moment that a wood drawer on wood guides just slides into its opening and fetches up against its stop, expelling a little puff of air from the cabinet, is a moment that provides much satisfaction.

TIP

Leave an inch between the bed slats to keep the mattress well aired and to avoid mildew.

Beyond the Footboard

With the drawers and the platform in, I had to decide what to do about the divider between the bed and the vanity. Here was where the self-imposed constraint of using the existing baseboard as the perimeter of the alcove began to bite. Because its height was determined by the slope of the ceiling, the mirror over the dressing table had to be as far forward as possible. But to bring it right up against the inside edge of the upper shoji track would eliminate the space required for a light above the mirror. And even that would put the top of the mirror at barely 6 ft. Temporarily derailed on the fast track, I tried to find other ways to light the mirror and kept coming back to the necessity of recessing a fluorescent fixture into the ceiling.

The fixture I used is a standard T-12 fluorescent fixture equipped with an Ultralume™ lamp. The lamp emits more lumens per watt than a standard cool-white lamp and has a higher Color Rendering Index, both important factors in getting an accurate reading on colors, like those at a makeup table.

Casting an even light across the face of the person standing at the mirror is important. So I put a narrow strip of mirror along the bottom edge of the large mirror, angled upward to bounce the light where it can fill in shadows.

Bookcase Wall

As for the partition between dressing table and bed, my fast-track conviction that it could not be frame and drywall held up better. My daughter wanted more bookshelves, and the foot of the bed was a logical place to put them (photo, right). I made the back of the bookcase out of ¾-in. birch plywood, which could be finished naturally on the book side and painted white on the dressing-table side to look like a wall.

To light the makeup table, I mounted a standard incandescent ceiling fixture in the space behind the mirror. On a playful impulse, I wired another of these lower on the sloped ceiling in the space behind the vanity drawer (the case for these has no back, so the fixture is easily accessible). Then, after carefully laying out the location of the bookshelf dividers and following a square-and-triangle motif suggested by the conjunction of the ceiling and the shelves, I jigsawed the holes in the birch ply and glued rice paper over them. This created little backlit rice-paper windows in the bookshelves. The dividers cover the edges of the paper. The only slight snag in this assembly is that the plywood thickness causes a shadow line, which can be seen where the backlighting travels at an angle through the window. If I'd thought of it in time, I could have easily eliminated the shadows by beveling these edges with a router.

The bedside reading light was more of a problem. Initially, I placed my standard ceiling fixture under the skylight as far down the slope of the ceiling as I could. I made a cardboard mock-up of the rice-paper shade that I had in mind to establish just how big it should be—the trade-offs being the height of the fixture, the size of the shade, and its

Bookcase wall. Shelves deep enough for paperbacks are affixed to a ¾-in. birch plywood panel between the bed and the vanity. The squares at the end of each shelf frame rice-paper windows that are backlit by bulbs behind the vanity drawers.

proximity to errant elbows. I thought I had a satisfactory balance, so I went ahead and made the lamp. But Genevieve put her elbow through it the first night she slept in the bed. I forgave her and accepted the lesson. The second reading light ended up above the head of the bed (photo, right).

What about the Shojis?

The shoji screens have yet to be made, and it now seems unlikely they ever will be. Although she was initially keen to have them, Genevieve now believes they would get in the way, and I agree with her. We analyzed the patterns of opening and closing that might be required during a typical day and night. It became clear that in spite of the desirability of drawing a curtain over the unmade bed by day and the unfinished homework by night, this teenager would rather live and sleep in one room—at least for the time being—than be bothered sliding screens to-and-fro all the time. A feeling of confinement was also a factor. Having tried out the bed myself one night when she was sleeping at a friend's house, I too felt I might want more distance between myself and any enclosing screen.

I admit that this was something of a blow to my vision of the room. What about the function of the skylight as a backlight for the shoji? What about the square-and-triangle motif I was going to incorporate into the shoji lattice? Ah, well, at least I hadn't made them already. And the grooves in the bottom track appear to work perfectly as 9-ft.-long pencil trays.

The rejected shojis and the difficulties I had with the makeup light and the height of the mirror were all results of my decision to keep the bed alcove within the area beyond the existing kneewall. If I had moved this line 6 in. to 12 in. into the room, I could have raised the upper shoji track a few inches,

Headboard. A reading light inspired by the bookcase's square-and-triangle motif lights up the headboard side of the bed. On the left, wooden tracks for shoji screens frame the alcove.

creating plenty of space to mount the mirror light, the reading light, and the shoji screens. The amount by which this would have reduced the size of the room would have been insignificant in relation to the space gained by building in the bed and the dressing table—a case of choosing the wrong existing condition to work from.

At least the client is satisfied. The project was completed during one of the long dry spells that Vancouver is famous for. Finally, one morning when the spider webs were glittering and the earth smelled refreshed and autumnal, Genevieve appeared downstairs with a beatific smile on her face. "It rained on my skylight last night," she said.

Tony Simmonds runs DOMUS, a design/build firm in Vancouver, British Columbia, Canada.

A Bookcase That Breaks the Rules

■ BY GARY M. KATZ

I remember the first bookcase I ever built. I didn't want to spend too much money on material, so I used AC fir plywood and had to sand the stuff until my hands hurt. I wanted the case to last forever, so I dadoed all the shelves into the sides— a time-consuming job, especially when my ¾-in. router bit was a little too wide for the ¾-in. plywood shelves.

I attached the face frame one piece at a time, which meant the joints weren't fastened together tightly, which was okay. But I made the shelves the same depth as the sides, so when I installed the nosing on the shelves, the face frame spread apart. Then, after all that, I had to finish the thing.

Since that first attempt and during my 30-plus years of finish carpentry, I've built a lot of bookcases—not just for clients, but for myself as well. Over that time I have learned a lot of valuable lessons about bookcase design and construction. I have some tips and methods as well as some misconceptions about building a better bookcase faster and easier.

Use Readily Available Materials

I've built bookcases from many different materials, from ¾-in. fir plywood to 2×12s. On occasion I've also used ¾-in. veneered MDF-core (medium-density fiberboard) sheet goods, but I've found that plywood is stronger, spans farther, and holds screws better than MDF. So I usually choose ¾-in. hardwood-veneer plywood for stain-grade work or ¾-in. birch-veneer plywood for paint-grade work.

I make stiles, rails, and nosings from solid stock to hide the edge grain of the plywood sides and shelves. I prefer poplar for paint-grade bookcases, and for stain-grade work, I use hardwood that matches the plywood veneer, in this case Honduras mahogany.

The bookcase back can be made from any ¼-in. sheet stock, but I opted for mahogany-veneer plywood to match the rest of the material. By the way, MDF core is fine for the bookcase back.

Challenging Bookcase Myths

Shelves don't have to be 12 in. deep; 9½ in. is deep enough for most books.

Dadoes aren't necessary for shelf support; screws are strong enough for bookshelves.

Adjustable shelves aren't necessary; no one changes shelf heights, so make them permanent.

Rabbeting the back into the sides is unnecessary and time-consuming.

Shopping List

One sheet of plywood yields all the parts for the main case plus the optional outer sides.

- One sheet: ¾-in. plywood (shelf, sides and top)
- Half-sheet: ¼-in. plywood (back)
- Four: ¾-in. by 2-in. by 60-in. stiles (face frames)
- Two: ¾-in. by 4½-in. by 36-in. rails (face frame)
- Five: ¾-in. by 1-in. by 36-in. nosings
- One: 1-in. by 1¾-in. by 48-in. crown molding

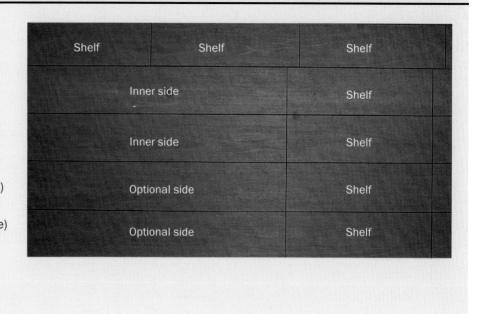

I keep costs to a minimum by using plywood for the shelves and sides and solid stock for the stiles and rails. The materials for this 32-in.-wide bookcase in Honduras mahogany (with the face-frame end panel) cost about $230 ("Shopping List," above). Paint-grade materials would cost about half that amount. Considering that the bookcase took me less than a day, I figure it could be built, finished, and installed for around $500.

Getting the Most from Material

I can get the shelves and sides for a single 32-in. unit easily out of a single sheet of plywood. I start by ripping the two finished outer sides (if both sides are exposed), and then the two narrower inner sides (bottom photos, p. 52). (The outer sides are ¼ in. wider to cover the edges of the back.) I make the height of the bookcase around 60 in. so that the cutoff pieces from each of the sides can be used for shelves. Enough material should be left over for the top or for shelves on additional bookcases.

I cut the material for the back with the grain running horizontally. That way, I get the backs for two bookcases out of one sheet of material. When all the bookcase pieces have been cut, I put nosing on the shelf stock, and I prefinish everything with a couple of coats of polyurethane.

Join Several Sections for a Modular Library

Lightweight and easy to carry, 32-in.-wide units can be fastened together on site for larger-capacity bookcases.

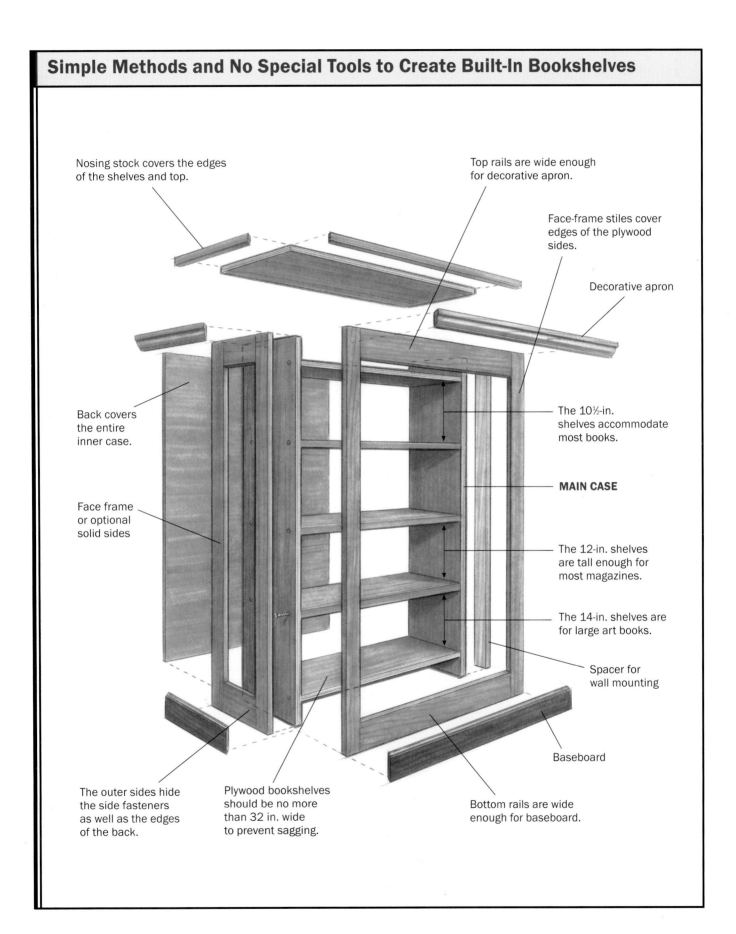

Nosing stock covers the edges of the shelves and top.

Top rails are wide enough for decorative apron.

Face-frame stiles cover edges of the plywood sides.

Decorative apron

Back covers the entire inner case.

The 10½-in. shelves accommodate most books.

MAIN CASE

Face frame or optional solid sides

The 12-in. shelves are tall enough for most magazines.

The 14-in. shelves are for large art books.

Spacer for wall mounting

Baseboard

The outer sides hide the side fasteners as well as the edges of the back.

Plywood bookshelves should be no more than 32 in. wide to prevent sagging.

Bottom rails are wide enough for baseboard.

Cutting Plywood Without a Tablesaw

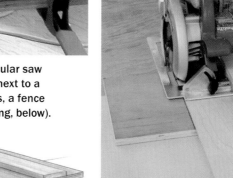

RIPPING JIG
Saw table rides on plywood base next to a guide strip.

Two tricks for straight cuts. Jigs can turn an ordinary circular saw into an accurate cutting tool. For rips, the saw table rides next to a guide strip (photo, above and drawing, right). For crosscuts, a fence squares the jig to the work to be cut (photo, right and drawing, below).

CROSSCUT JIG
Fence screwed to bottom edge squares the jig to the work.

Turning the plywood into bookcase parts. First, rip the sheet of plywood into lengthwise strips. A table saw makes these cuts straight and even (left). Next, a chopsaw makes quick work of cutting the strips to length for the shelves, sides and top (above).

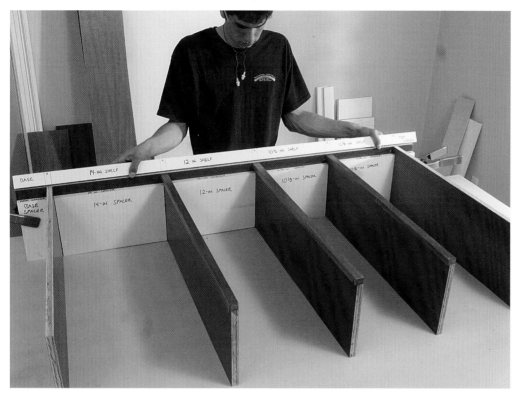

A story stick records the vertical layout of the bookcase to minimize measurement errors and to keep the layout consistent with multiple cases. Measured from the story stick, 1×8 spacer blocks clamp to the sides and keep the shelves properly spaced and positioned during assembly.

Assembly Requires No Special Tools

Story sticks can be helpful in just about every phase of home construction. They usually are made from a length of 1× material, and they're used to lay out and record all the pertinent measurements for a particular project. For a bookcase story stick, I mark the location of each shelf as well as the width of the top and bottom face-frame rails based on the trim details I plan to use.

I use the story stick to lay out temporary spacers that I cut to set the shelf positions exactly on both sides of the bookcase (photo, above). The story stick and spacers also can be used to make additional bookcases identical to the first if necessary.

I attach the shelves to the sides with 1⅜-in. drywall screws driven through predrilled holes. To avoid "shiners" (screws or nails that miss their mark), I trace a pencil line with a square at each shelf

location before driving any fasteners. Before drilling and driving the screws, it helps to tack the shelves in place with 18-ga. brads (photos, right).

Back Squares the Bookcase

I secure the back to the assembled case with glue and staples (top photos, p. 54), although small nails or screws work, too. Once the glue is spread, the back must be set carefully to keep the glue from oozing. If glue does squeeze out, it's easily cleaned off the prefinished pieces with a wet cloth.

I first attach the back along just one side. Although I cut the back square, I double-check by racking the case until the diagonal measurements are exactly the same. Again to avoid shiners, I use a straightedge to draw the shelf locations across the back before stapling it home.

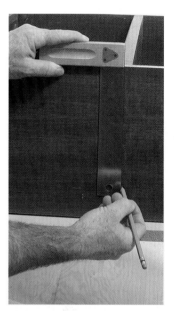

Pinpoint the screws. Once the shelves are positioned, a line is drawn to center the fasteners (above). The shelf is held against a spacer while the screws are driven through predrilled holes (below).

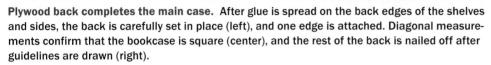

Plywood back completes the main case. After glue is spread on the back edges of the shelves and sides, the back is carefully set in place (left), and one edge is attached. Diagonal measurements confirm that the bookcase is square (center), and the rest of the back is nailed off after guidelines are drawn (right).

Hidden Fasteners Hold Bookcase to the Wall

If the bookcase is going in a corner, I install a 1× spacer block on the end panel to keep the stile overhangs similar. Then I set the unit in position and shim it plumb.

One of the advantages of this bookcase system is that all the fasteners are hidden. The ¾-in. plywood sides have plenty of strength, so I drive a screw through the side that abuts the wall and into a wall-framing member. However, the ¼-in. plywood back isn't strong enough by itself to fasten to the wall. So I drive screws through angled holes in the top shelf. Another option would be to add a reinforcing cleat across the back of the top space. A third attachment point is through the side with a screw into either a stud or the bottom sill plate.

Face Frame Covers Raw Edges

The end panels go on next to hide all the fasteners in that side. The next step is installing the face frame, which covers the

Spacer keeps trim consistent. A 1x spacer is screwed to the side that goes against the wall (inset photo, far left), and shims make the bookcase level and plumb (photo, left).

Three Places to Hide Screws

1 Drive screws through the sides into wall studs.

2 Drive angled screws through the top shelf into wall studs.

3 Drive angled screws through the side and into studs or sill plates.

Fancy finish. To give the bookcase an Arts and Crafts look, a stile-and-rail frame covers the shelf screws (below left). Screws secure it from the inside, and another preassembled frame hides its edges as well as the edges of the plywood (below right).

Sources

Kreg
(800) 447-8638
www.kregtool.com

unfinished edges of the plywood. The face frame can be installed one piece at a time on the case, but I prefer to assemble the face frame using pocket screws. Pocket screws ensure tight-fitting and flush joinery. Then glue and a few 18-ga. brad nails are all that's necessary to secure the face frame to the carcase.

Next, I apply nosing to a piece of ¾-in. plywood for the top, and I glue and nail it in place. I like to put an apron under the edge of the top. A simple piece of 1×2 works well, but other moldings that are either made in a factory or made in my shop can be used to gussy up the design.

Gary M. Katz is a carpenter and writer living in Tarzana, California. He is the author of The Doorhanger's Handbook, *published by The Taunton Press, Inc., in 1998).*

Baseboard completes the built-in look. A nosing strip on the top hides the edge of factory-made crown molding, and a matching baseboard finishes the bottom.

Plywood panels keep costs down. Minimal and clean-looking, side panels cut from the same sheet of plywood hide the screws that hold the shelves (right). The preassembled face frame goes on in one piece to cover the edges of the plywood (far right).

Modest and modern. A square-edge apron under the top adds a sleek, simple touch.

Designing Built-Ins

■ BY LOUIS MACKALL

We live in the age of stuff. As an architect who specializes in kitchens and other interiors, I get to restore some semblance of order. But early in my architectural practice, I grew frustrated by how far removed architects typically are from the materials and processes of building. I remember saying to friends that I felt like a guy shouting instructions to a couple on their first date. I was on the outside looking in, missing out on all the fun; that's why I became a cabinetmaker.

Before it's cut, a board is like a blank piece of paper, gently asking what I will make with it that would be more desirable than the board itself. As an architect, I find the voice of the board hard to hear. As a cabinetmaker, I have heard its simple question every time I approached the table saw.

In the following article, I'll explain an informal set of rules that I use to design everything from kitchen cabinets to built-in beds. Along the way, I'll illustrate these rules with photos of projects that I've designed.

Hardly your typical wall cabinet. A glass door with a sandblasted design and interior lights begins to set this cabinet apart. But note, too, how the cabinet combines with the arched valance over the window to create a soffit for recessed lights and an opportunity for display shelves.

Recessed built-ins don't seem to rob space from the room. A secondary wall of MDF around this living room allows the display cases, cabinets, drapery pockets, and valances to be recessed. The varying depth of the recess is revealed at the doorways.

Rule 1: Add to the Room with Its Cabinetwork, Don't Take Away.

Built-in cabinets should be just that—built in—not placed around the room like furniture. To rid myself of preconceptions, often the first thing I do is fill the room with foam.

I'm speaking metaphorically, of course. But as the foam sets up, my mind squirms under the inevitable "What now, fella?" After the foam is hard, I proceed to carve out the room, one eye on the list of things the clients have requested and the other eye on the spaces being created between them.

In the living room shown on the facing page, I needed to accommodate a variety of things: my client's pre-Columbian art collection; pockets for the heavy drapes and a valance for each of the exterior sliding doors; the mantelpiece; and extending all around the room, a clerestory with light behind frosted glass to extend the sense of space and to convey a feeling of warmth out to the edge of the room. Inside the room's framing, I established a secondary wall that defines the apparent limit of the room. The depth behind the wall varies. It is shallow for the curtains, but deeper for the cabinets. Openings through that wall create various niches, clearly subordinate to the room itself, minivistas to ancillary worlds. Done this way, the built-ins have become one element, and the room's stuff is an addition to the space of the room, not a subtraction from it.

Rule 2: Where Possible, Light Cabinets from Inside.

The impact of upper cabinets is always much more significant than that of the lowers, and deserves more attention. Where possible, I use glass doors—either frosted or clear—for upper cabinets. The glass allows a brightness that a solid door does not. Glass also allows me to put lights inside the cabinets.

In the evening, with all the other lights off, the pockets of light from cabinet interiors create a pleasant warmth not unlike twilight, or the first, predawn rays of the sun. My "sunset" theory says that light coming from the side is much more enjoyable than light coming from the ceiling.

In the kitchen shown on p. 57, there is lighting around the inside edges of the upper cabinet-door openings, as well as around the opening that frames the sink window. The light used here, and in the living room mentioned earlier, is called Bendalite. It is a ½-in. dia. clear-rubber tubing with small incandescent lights inside (photo, right). It comes in 30-ft. lengths, lists for about $6 per ft. and is rated for 25,000 hours. It is easily run around irregular openings and through ⅜-in. holes drilled in the cabinet walls as a continuous hose of light.

Rule 3: Organize Elements by Layers or Planes, with Special Attention to Edges.

All the elements of a building are important, but windows more so because they are our connection to nature. How the window is presented can determine whether people are drawn to it and, ultimately, how much they like the view. In the kitchen shown on p. 57, each window is framed by an arched, beaded proscenium. Contained in the frame are the lights and display shelves. All these things are detailed to give proper space for the window casing and sill. The cabinets on each side reinforce the frame.

Flexible lighting is easy to install. Sliding-glass panels circle above the living room's built-ins. Routed inside, strings of tiny incandescent lights in flexible tubes create a warm glow.

Like a lighthouse in the middle of the kitchen. A freestanding kitchen workstation (right) helps to break up a large space, shortening the cook's travel time to and from the stove and creating a book-filled backstop for a sitting area on the other side (bottom left, facing page).

Rule 4: When Necessary, Use Built-Ins as Room Dividers.

Most kitchens (most homes, in fact) are designed and fabricated on homogenized rules of value and utility, which work okay for the majority of users and are efficient for a factory's production of cabinets (or closet doors or shelving units or whatever). Unfortunately, somewhere along the way the poetry evaporates, which is painful to someone for whom objects have meaning.

One of the homogenized rules for kitchens is that except for the occasional peninsula or island, cabinets go against the wall. But the room for the kitchen shown above right was too big, so I used a big kitchen island as a way of breaking down the scale of the room and shortening the walk between stove and sink. As one piece of furniture, this island combines all the basic functions of a working kitchen, except the refrigerator, while adding a significant expanse of bookshelf for the sitting area at the fireplace (bottom left photo, facing page).

Note the upper cabinet's glass doors. All of the glass doors are the same size and lit from above, creating a kind of lighthouse in the middle of the room. Holding the top short of the ceiling helps to make the kitchen ceiling seem just a bit higher. Behind a sliding door at the back of the sink, there is a TV set. Above, there is a framed opening into the sitting area looking onto the fireplace.

On another project, I used the same principle—built-in as room divider—to create a headboard with built-in nightstands, lighting and bookshelves (top photo, p. 62). Planted in front of the closet and bathroom doors, the headboard also serves as a kind of privacy screen for dressing and on the back side made a place for sock and sweater drawers (bottom photo, p. 62).

Curved shelves in a rectangular case. Building out one wall in the kitchen to recess the refrigerator and stove created this deep passageway between the dining room and the kitchen (top and left). It also created an opportunity for shelves that overhang their casework and sport subtle curves in their leading edges (above).

A built-in headboard becomes a room divider. With integral nightstands, bookshelves, and lighting, this headboard makes a cozy spot for curling up with a book. The chamfered curves that frame the opening of the headboard were cut into medium-density fiberboard, which was then painted.

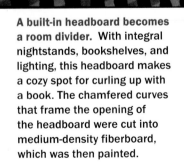

The back of the headboard is a dresser. Outfitted with a mirror and a bank of drawers, the back of the headboard serves as a built-in dresser. Positioned as it is, the headboard also functions as a privacy screen, creating a dressing area between the closets and the rest of the room.

Rule 5: Use Curves Whenever You Can.

Curves soften life. Straight lines are the easy way out, like those kitchen cabinets that arrive in cardboard boxes; they're the visual equivalent of a dial tone. Curves are liberating.

The prettiest curves are entirely free form. Butterflies in space. Next come ellipses, more prescribed than butterflies, but at least the radii vary. Finally, there are arcs, still head and shoulders above straight lines. Any curve puts life into an edge.

The curve at the bottom of the upper cabinet in the kitchen that's shown below skips over Curves 101 and goes directly to a post-graduate school of arc, thanks to a couple of twists. First, the radii of the various arcs

change as they begin and end, which conveys an intimacy that an arc with constant curvature does not. It guides the eye on a little tour, or dance, slowing down for corners, endings and transitions. It's a mini roller coaster.

Second, the curve changes direction. At the right side of the cabinet, the curve coming across the face burps a couple of times, then turns 90° and loops back, and then goes down to the plane of the cabinets below, stopping just above the counter's surface. Looking at the counter, your eye sees the curves and also the space defined by the curves. Their interaction creates a three-dimensional event. As I remember, it was hours on the floor of our shop with my pencil and a shrinking eraser before I was satisfied with what I had done.

When a curve becomes an event. The voluptuous curves across the bottom of this wall cabinet do nothing for its ability to hold dishes, and those curves hardly simplified the cabinet's construction. But they animate the cabinetry in a delightful way. Notice how the muntin design in the cabinet matches that of the window.

A wide hallway becomes a home office. This built-in desk illustrates many of the author's "rules." The upper cabinets with their frosted glass and decorative muntins get more attention than the base cabinets. Elegant curves grace the ends of the desk and the apron below the paper cubbies.

Simple stops retain the glass and the decorative muntins. The cabinet doors over the desk are simple frames with a rabbeted inside edge. Both the muntins and the glass are laid into the rabbet and retained by a simple wooden turn button.

Rule 6: Exploit Muntin Patterns to Integrate Cabinets or Simply to Animate Them.

It's important that the various elements of a design converse with each other. In the kitchen shown on p. 63, the pattern of the muntins in the cabinet doors reflects those of the adjacent window. The scale of the upper cabinet is large. The width is more than usual, and the doors are hung on standard brass hinges. All of these style cues complement the age of the house and also help to distinguish this cabinetwork from the stuff that generally arrives at the house in little cardboard boxes.

The cabinets over the desk shown on the facing page continue our long tradition of fooling around with muntin patterns as fretwork. Quite a while ago, I developed a basic approach for thin muntins whereby they are made separately as a complete frame element. The door is made with a simple rabbeted opening. With the door face down on a bench, the fretwork is laid in place, glass laid on that as one piece, and stops applied to hold the sandwich in place. Or in some places—a kitchen, for instance—we'll substitute rotating tabs for the stops (bottom photo, facing page) so that the glass can be taken out easily when it's time for cleaning.

Rule 7: Let Shelves Overlap Their Casework When Appropriate.

On the same built-in desk, the bottoms of the paper cubbies have a separate piece of thin pine that was laid in after the rest had been painted. This shelf laps out over each side of the opening, creating tiny sills. These tiny sills are a kind of whimsical miniversion of something I do whenever possible, which is to let the shelves overlap their casework (bottom right photo, p. 61). Basically, this rule is an excuse for more curves because letting shelves slip their bonds makes it easy to bandsaw a subtle curve to the shelves' leading edges.

Rule 8: Paint Almost Everything.

By now, you've probably noticed that most of my stuff is painted and usually white. Forms read more clearly when they're painted than when they're left as natural wood, and I've come to appreciate that clarity. In addition, painted work integrates more easily with the walls of a room and is a pleasant foil for the natural materials frequently used for counters and floors.

Rule 9: Integrate Everything and Think Through All the Details.

For each of these projects, I started out from the big picture—"we need a kitchen"—and followed the thought down to the tiniest details—"where can I hide the light switches?"—for these items are the dessert. Wherever possible, I integrate all the requirements into one and create a hierarchy of planes to house all the elements, almost as strata in an archaeological dig or files on your computer's hard drive. I don't always get it right, but I almost always get to try it again. And the second time, I remember that the bullnose on the granite counter should stop short and revert to square just shy of the square-edge commercial range.

The pleasure all of this provides returns twice. The first is private, for me while I draw. The second is public after it is built, and with any luck continues for the life of the house.

Louis Mackall practices architecture in Guilford, Connecticut, and is the president and co-founder of an architectural-millwork company called Breakfast Woodworks.

Sources

Bendalite
905 G Street
Hampton, VA 23661
(800) 448-2014
www.bendalite.com

A Pair of Built-In Hutches

■ BY KEVIN LUDDY

One thing I've learned from working as a carpenter for 20 years is that home-owners are uncharted waters. When I met with the owners of this house about build-ing a pair of hutches, two things became apparent. First, even though they wanted to keep the price down, they wanted a quality job. This was good.

But the second thing was that these hutches had to fit into two alcoves that flanked an existing fireplace, and these alcoves were ridiculously out of square. And although they were nominally the same size, the two spaces were more than 2 in. different in both height and width, a fact that threat-ened thing number one. Another threat to the economics of the project was that the widths of "about 4 ft." (as I had been told on the phone) were actually 49 in. and 52 in., not great for efficient use of sheet goods.

A Dry Run in the Shop

Each of the paint-grade hutches would con-sist of a base section with cabinet doors and a countertop, and an upper section with open, adjustable shelves. The cases and shelves would be ¾-in. birch veneer-core plywood, and the doors and moldings would be solid wood. To save material, I planned to use the existing walls in the two alcoves as the backs of the hutches.

I prefabricated all the parts for the hutches in my shop. I also made the raised-panel doors for the bases and the trim pieces that would hide the variations in the walls and ceiling. I preassembled the bases and shelf units in the shop to make sure everything would go together easily, and then the parts were sanded and given a good coat of primer.

Lots of Shims and Scribing

When I arrived at the site, a finished house, the first things I unpacked were drop cloths. I reacquainted myself with the site dimensions and peculiarities and then located the framing behind the drywall for attaching the case sides.

Faced with spaces that weren't close to square, this carpenter judiciously applied trim and careful scribing to hide the worst of it.

The wall in one of the alcoves was out of plumb almost ⅜ in. over the height of the lower cabinets.

With the existing walls acting as the backs of the cases, the first pieces I installed were the sides of the lower cabinets. At the shop, I'd notched each side for the toe kick. The floor in one of the alcoves was fairly level, so I plumbed the sides in place and scribed them to the back wall, keeping the top edges level (more on scribing later). When I was satisfied with the fit, I glued and nailed the bottom shelf to the two sides (top left photo, below). Then I centered the assembly in the opening, shimmed each side plumb and shot nails to hold each side in place (bottom left photo, below).

The floor of the other alcove was not level, so before scribing the back edges, I had to shim the lower of the two sides up from the floor. The height differences were to be hidden by a custom-fit toe kick installed later. Next, I screwed a cleat to the back wall on each side to support the back edge of the countertops (top right photo, below).

When fitting the countertops, I slightly undercut the back 12 in. on both sides of

After the sides of the lower cabinets had been fit to the back wall, the bottom shelf was glued and nailed to cleats on the sides.

After the lower-cabinet sides and shelves are put in, a 1x cleat is installed to support the back edge of the countertop.

Before continuing with the construction, the author installs rails under the front edges of the countertops for added strength.

With the existing drywall forming the back wall of the new hutch, the sides with the bottom shelf attached are centered in the opening and shimmed plumb.

To scribe variations of less than ⅛ in., the author lets a pencil ride on the wall, its point recording the scribe line.

each top (which would be hidden under the upper cases) to allow for easier fitting. The tops were then scribe-fitted to the back wall and nailed in place. Before going any further, I glued and nailed the rails under the countertops to give them added strength during construction (bottom right photo, facing page).

Fitting the Top Pieces is a Grind

Before fitting the sides for the upper-shelf sections, I tacked shims to the drywall, using a level against a straightedge to keep the shims in a plumb line (photo, left). The sides were the toughest pieces to fit because they were captured on three sides by the ceiling, the back wall, and the countertop, and all these joints were visible. Starting with the tallest side (so that I could use it in a shorter spot if I screwed up), I scribed each side piece to the textured ceiling first.

Then I set each piece in place at a slight angle because of the tight top-to-bottom fit and pencil-scribed each piece to the back wall (photo, above). For this scribe of less than ⅛ in., I held a short pencil to the wall and let the point ride on the piece to be fit.

Using a straightedge and a level, the author tacks shims to the wall to create a plumb line before the sides are fit.

An electric grinder creates a bit of dust, but the grinder is easy to control and allows a feather edge to the scribe line.

To make the sides perfectly parallel, a shelf is set in place, and the shims are adjusted to the shelf.

I've seen many different tools for cutting scribe lines: jigsaws, hand planes, rasps. For most scribes, though, I prefer a small electric grinder (photo, left). It may create a little more dust, but I think a grinder is easier to control and to feather to the line.

The side pieces were then tacked in place. To keep them parallel while they were being nailed, I rested a shelf on its supports between the sides and shimmed as needed (photo, below).

Hiding a Multitude of Shims

I'd now reached the magic portion of the show, installing the molding and face frames to finish the piece and to camouflage all the differences in height and width. I started at the top with the crown molding and the backer board installed underneath.

Much to my amazement and luck, the ceiling above the left-hand hutch was level (all the error was in the floor). So the crown backer and molding went in quickly and easily (top photo, facing page). The installation gave me a reveal of about 1¾ in. of backer board below the crown. My luck ran short on the right side, where I needed to make up almost ½ in. in 4 ft. I decided to hide this difference in degrees. First, I applied the backer board ³⁄₁₆ in. out of level and then nailed the bottom edge of the crown, varying the reveal from 1¾ in. to 1⅞ in. To hide the last ⅛ in. or so, I twisted the crown into position, making it slightly taller on one side than the other. When I stepped back from the hutch, I was impressed by how even the detail looked after all that fussing.

My bad luck continued for the upper face-frame stiles. I thought a 1⅜-in. width would give me plenty of stock to scribe to the wall and extend beyond the shelf sides by at least ⅜ in. But because I had to shim the ¾-in.-thick sides off the wall by as much as ½ in., the face frames barely covered the sides. I would have to make new face frames.

At my urging, the homeowners had decided on fluted stiles for the lower cabinets. The walls on the left side had little taper, so the fluted stiles went in with little tweaking. I biscuited the tops of the stiles into the rail and then glued and nailed them to the case sides and bottom.

I had made wider stiles for the right-hand cabinet, and I needed every bit to make up for more than ⅜-in. taper in the walls. Scribing these stiles took a bit more care so that I would be left with the exact finished opening for the doors when I was done. I made a couple of test fits and then biscuited, glued, and nailed the stiles in place. After fitting the doors, I called it a day.

Face Frames Stabilize the Shelf Sides

The next day, I returned with the wider upper face-frame stiles. Increasing their width to 1¾ in. allowed me to scribe them to the wall (photo, below left) with plenty of extra to hide the ends of the shelves. To scribe a taper of this magnitude, I grabbed my 99¢ compass-style dividers. I glued and nailed the stiles in place and put construction adhesive on the wall edge of the stiles to help stabilize the sides (photo, below right). I then gave the doors a final adjustment, filled the nail holes, and touched up the primer.

After the painter put on the finishing touches, I got a call from the clients saying that they loved the way the hutches turned out. The scribing had hidden the worst of the variations, and the trim details pull the eye away from the rest. The homeowners had picked out knobs for the doors, and I promised to put them in next time I was in the neighborhood.

Kevin Luddy runs Keltic Woodworking, a custom-carpentry and cabinetry company in Wellfleet, Massachusetts.

The first trim to be installed is the crown backer and molding, which is adjusted to disguise a ceiling that is out of level.

For big irregularities such as the ones here, inexpensive but invaluable compass-style dividers transfer the wall's contour to the piece being fitted.

To help stabilize the sides of the shelf sections, the face-frame edges are coated with construction adhesive before being nailed to the shelf sides.

Building a Lazy-Susan Cabinet

■ BY REX ALEXANDER

I have no idea who dreamed up the lazy Susan, but I bet a Hoosier had something to do with it. Indiana, after all, is the birthplace of the famed Hoosier cabinet, and Indiana is where I came across a lazy-Susan corner cabinet for the first time. I had just started working as a cabinetmaker, and I landed a kitchen-remodeling job that included the removal of every cabinet but one, a corner cabinet that housed two lazy Susans. It was a primitive affair that probably had been built on site, but the owner loved it.

Made of standard lumberyard plywood, the cabinet took up a 36-in.-sq. space with a cutout in the corner. Both the bottom and center shelves supported 32-in. rounds of plywood that revolved on low-profile bearings. Even though I've upgraded the materials and hardware and added solid-wood edging to the carousels, my lazy-Susan cabinets are virtually the same (photos, facing page). There wasn't much in that design to improve.

These cabinets hold a lot, and the 1,000-lb. capacity of the bearings means they work smoothly, even when the carousels are full of appliances or canned food. These cabinets eliminate wasted space in a corner, and they feature doors that open out of the way for easy access. Although I can buy factory-made lazy Susans, I think there are advantages to making these cabinets myself. My carousels are more attractive, operate more smoothly, and are a lot stronger than those flimsy trays revolving around a center pole. I can make mine any size. The 12-in. bearings of galvanized steel I use are inexpensive (about $7 each) and readily available from woodworking-supply dealers and catalogs.

Start by Cutting Out Pieces for the Cabinet Box

Although I stick with ¾-in. melamine for the cabinet box, I use ¾-in. hardwood plywood for the carousel pieces because it's less likely to chip. To cut melamine cabinet pieces to size on a table saw, I use a Freud LU98, an 80-tooth blade with a triple-chip grind. For ripping plywood, I use a thin-kerf rip blade. I don't get fancy with joinery on these cabinets. A simple dado joint cut on the table saw works fine. I use zero-clearance table-saw inserts with all these blades.

Two ways to hang the doors. When clearance is not a problem, the author hangs doors on 165° cup hinges so that they meet at the center of the cabinet (photo, left). If the cabinet is next to an appliance (photo, above right), the two doors are joined by a piano hinge at the center and hung so that they swing away from the obstruction.

After squaring and cutting the top, the bottom, and the shelf panels, I mark out 90° corner cuts on each piece. I start these cuts on a table saw (photo, left), but I'm careful not to cut too far and finish the cuts with a jigsaw (photo, bottom left). Bosch® makes a 4-in., 10-tpi jigsaw blade (model T101BR) with its teeth facing down instead of up as in a standard blade. It reduces splintering in wood and chipping in melamine.

To eliminate wasted interior space in the back of the cabinet and to make installation easier, I cut a 10-in. piece off the top, bottom, and shelf pieces (photo, below right). This cut is easily done on the table saw with the points up against the fence.

After the pieces for the cabinet have been cut to size, I then build the carousels for the lazy Susan. The 32-in.-dia. circles are cut from ¾-in. maple plywood. I use a plunge router and a solid-carbide spiral bit to make them.

Stop cut short of the line. Cabinet top, bottom, and shelf pieces are cut from ¾-in. melamine. To avoid overcutting the 90° corner, the author stops the saw before the blade reaches the layout line.

Finish with a jigsaw. With the workpiece on a pair of sawhorses, the author completes the cut with a jigsaw. A blade with downward-pointing teeth reduces splintering in the brittle top layer of melamine.

Cut off the cabinet's back corner. To make installation easier and to eliminate wasted interior space, the author nips off the back corner of the cabinet top, bottom, and shelf. This table-saw maneuver is safe, providing the fence is long enough. If not, clamp or screw an auxiliary piece to the stock fence.

Carousel Pieces Are Made with a Template or a Trammel

Because I make lazy-Susan cabinets frequently, I have a 32-in.-dia. plywood template for cutting out the circular carousel pieces. I screw down the template to the plywood (photo, below) and make a half-dozen or so shallow cuts around the edge to produce a carousel piece (photo, bottom right). If you will need a template only rarely, it may make more sense to use a trammel of ¼-in. plywood to make the carousel pieces. The trammel should be about 20 in. long and the width of your router base (photo, right).

To mark the 90° cutouts, I place the carousel pieces on the top, bottom, or shelf piece, adjusting them so that they

Making a Perfect Circle

A trammel made from ¼-in. plywood that pivots on an 8d nail is a good substitute for a circular template. To cut a 32-in.-dia. carousel, set up the router so that the distance between the nail and the inside of the router bit is 16 in.

A template makes the work go faster. To make plywood carousel pieces, the author starts with a 32-in.-dia. plywood template that is screwed to ¾-in. maple plywood. If you don't want to make a template, try a trammel attached to the router.

Use a guide bushing and light passes. With a solid-carbide spiral bit and a guide bushing, a router makes a clean cut in maple plywood around this reusable template. Cuts should be light—about ⅛ in. per pass.

are evenly spaced, and then mark the cut from below (photo, left). I use a jigsaw to make the pie-shaped cut (photo, below) and clean up the cut with a low-angle block plane and a chisel.

Plywood carousels are banded with a thin strip of wood. This edging piece must be high enough to keep the contents of the lazy Susan from falling off as it is rotated, and thick and low enough to resist splitting if it is bumped. After a lot of experimentation, I now use clear, straight-grained maple $7/32$ in. thick and 1½ in. wide. A piece 88½ in. long gives me enough to wrap around a 32-in.-dia. carousel with a little bit to spare. A band clamp with a rapid-action ratchet, available from Woodcraft Supply, works well for gluing (top photo, facing page). I apply a bead of aliphatic (yellow) glue to both the plywood and the maple

Mark and cut the carousel. The author centers a plywood carousel on a cabinet top, (bottom), or shelf piece. Before marking it for the 90° cutout, he moves the carousel toward the front of the cabinet ¼ in. to compensate for hardwood edging. A splinter-free blade is a good choice for taking out the pie-shaped piece on the carousel. The author finishes with a low-angle block plane and a chisel for a clean edge (top).

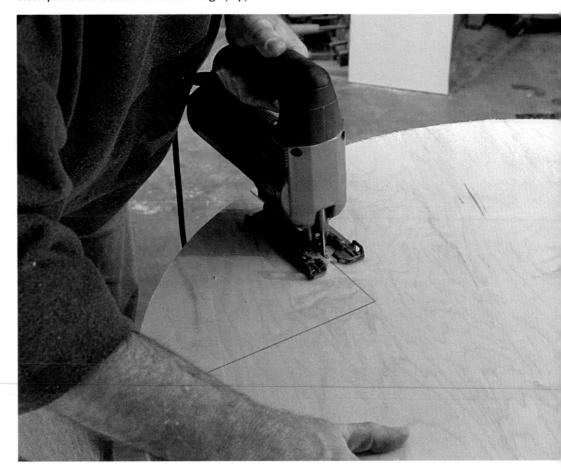

edging. C-clamps or spring clamps can be used to hold the edging in place while you put the nylon webbing in place. Glue squeeze-out should be cleaned up with a damp cloth.

Finish Up the Edging, and Attach the Bearings to the Carousels

I leave the band clamp on for 24 hours to make sure that the glue bond is good. Then I square the ends of the edging, cutting them close to flush with a dovetail saw and trimming them flush with a low-angle block plane. The two remaining pieces of the edging can then be glued and pin-nailed to the 90° cutout. I think that it's a good idea to nail the outside corner where the edging meets with a ¾-in. brad as a further guard against separation (photo, below).

For added strength, I apply preglued veneer to the outside edges of the carousels. The veneer covers and strengthens the butt joints, making them look mitered, and it gives the edging a full ¼-in. thickness. I apply it with a household iron set between "cotton" and "wool," and press it down with a J-roller. It can be trimmed with a utility knife.

Carousels should be finished before they are mounted on the waiting cabinet parts. I apply as many as six coats of precatalyzed lacquer, a tough and wear-resistant top coat. Precatalyzed lacquer is a highly flammable, solvent-based finish that should be applied in a spray booth. Whenever possible, I spray outside. If you don't want to bother with this kind of finish, I'd suggest Enduro Poly™, a water-based finish that can be brushed on.

Add the edging. Straight-grained maple 1½ in. high is glued to the perimeter of the carousels to create a lip that prevents contents from sliding off as shelves are turned. The author uses yellow glue and a web clamp to attach the edging.

Outside corner is reinforced with brads. A potential trouble spot is the outside corner of the carousel, where edging pieces meet. Here, the author adds some ¾-in. brads. Drilling pilot holes will prevent splitting.

Sources

Woodcraft Supply
P.O. Box 1686
Parkersburg, WV 26102
(800) 535-4482
www.woodcraft.com

Compliant Spray Systems
1242 Puerta Del Sol
San Clemente, CA 92673
(800) 696-0615
www.compliantspraysystems.com
Enduro Poly

TIP

For added strength, apply preglued veneer to the outside edges of the carousels with a household iron and a J-roller. Trim with a utility knife.

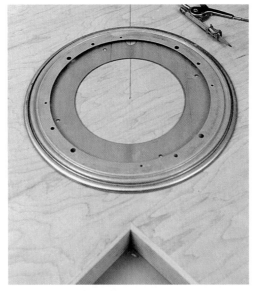

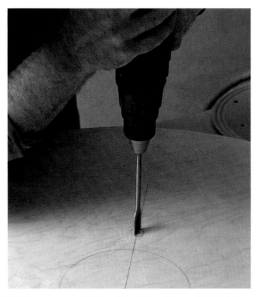

Lay out hardware location. With the carousel upside down, the author centers the lazy-Susan hardware. A line drawn through the centerpoint is used to locate a hole for mounting the carousel.

Drill an access hole. Once hardware has been attached to the bottom of the carousel, the only way to screw that assembly to the cabinet will be through a ¾-in. access hole in the carousel.

Attach hardware to bottom of carousel. After the carousels have been finished, the author flips them upside down on a padded bench and adds the lazy-Susan hardware.

Now mount carousels on cabinet parts. Working through the access hole, the author attaches a carousel to the cabinet shelf. After inserting a screw, the author revolves the carousel until the next screw hole appears below the access hole.

Assembling the Carousels and Completing the Cabinet

Installing the lazy-Susan bearing is a two-step process: Start by attaching the hardware to the carousel, then screw that assembly to the cabinet shelf or cabinet bottom. When both bearings are attached, the cabinet can be assembled. The bearings I use have a 6½-in. hole in the center, plus a ¾-in. access hole in the bottom flanges (top left photo, facing page). With a compass at the center of the carousel (use the hole left from the template or trammel), I draw a 6¼-in. circle. Then I draw a straight line from the point of the 90° opening to the carousel's center-point. I line up the bearing with the circle and set the access hole over and centered on the bisecting line. I mark the access hole with a pencil and drill halfway through with a ¾-in. spade bit (top right photo, facing page). Finish the drilling from the other side.

After screwing the bearing to the bottom of the carousel (bottom photo, facing page), I flip over the piece and place it onto the bottom cabinet panel. I make sure the

Putting it all together. Dado joinery is simple and strong, and makes for speedy cabinet assembly once all the pieces are ready. The author will cover raw particleboard edges in the cutout by ironing on preglued polyester tape.

carousel spins evenly, and then screw it into place through the access hole (top left photo), turning the carousel until the access hole exposes an empty screw hole below. After that, assembling the cabinet goes quickly. I use yellow glue on the dado joints and fasten the pieces with narrow-crown 1-in. staples in a pneumatic nailer (photo, above). When I'm gluing on the back, which is a melamine-to-melamine joint, I make sure to use a glue made specifically for this material.

I finish by ironing on preglued polyester tape to the 90° cabinet opening and trimming any excess with a utility knife. Doors are hung on cup hinges.

Rex Alexander, a frequent contributor to Fine Homebuilding *and* Fine Woodworking, *builds cabinetry and furniture from his home shop in Brethren, Michigan.*

A Built-In Hardwood Hutch

■ BY STEPHEN WINCHESTER

I love an old house. Working on one makes me appreciate the skill of the carpenters who came before me. It's amazing to see the level of craftsmanship the old-timers attained using only hand tools—especially in their trimwork. I recently renovated an early 1800s farmhouse in New Hampshire that had some beautiful chestnut trim. I got the chance to match this woodwork when I added a family room with a built-in hutch.

I made the new family room by removing a wall between two small rooms. There was a closet in each room, one on both sides of the wall, and when the wall came down, the closet area was a natural location for the built-in hutch. Built-ins ought to look good and last a long time, so this hutch was built of solid hardwood and designed to accommodate wood movement (drawings, pp. 82–83). But before I started building, I straightened and leveled the closet area.

Roughing In the Hutch

New studs on the left and right made the sidewalls plumb and straight, but there wasn't room on the back wall for new studs. So I straightened the back wall with shims and 1×3 strapping (bottom photo, facing page). At the bottom I tacked a 1×3 across the old wall and into the old studs. Placing a straightedge on the 1×3, I tucked some shims behind the low spots to bring them out to the straightedge. Next I tacked a 1×3 to the top, again shimming it straight. Then I tacked on more horizontal 1×3s 16 in. o. c. Moving from left to right, I held the straightedge vertically, against the top and bottom strapping, and shimmed the intermediate strapping out to the straight-edge. The wall was straight when all the pieces of strapping were even with each other.

The hutch rests on a 2×4 base; I installed it level by shimming the low end and nailing it to the new 2×4 walls on each side. With the new level base, I didn't have to scribe the cabinet sides and back to the floor, which had a big hump in it.

Ash matches. A new ash hutch built into an old closet looks like the chestnut woodwork of the original room. The hutch was finished with two coats of Minwax® Polyshades®—half maple and half walnut—followed by a slightly thinned top coat.

Chestnut Substitute

Chestnut was once used for almost everything in a house, from sheathing to door and window frames to trim. But during the first part of this century, a blight wiped out almost every American chestnut tree. Today, you can get salvaged chestnut from old buildings or get it resawn from beams or sheathing, but it's expensive. I chose white ash instead, which has about the same grain pattern and texture as the chestnut woodwork on this job. But ash is hard, so it's more difficult to work than chestnut.

The opening. New studs frame the sides, but the back wall of this former closet was straightened with 1×3s and shims.

Gluing Up Wide Boards

The cabinet floor and the counter were glued up out of several boards, as were the wide shelves for the bottom cabinet. To joint and join the boards in one step, I used a glue-joint cutter in the shaper. (Jointing is the process of straightening a board's edge or face and is typically done with a jointing plane or with an electronic jointer. Joining is the process of connecting two boards.) The glue-joint cutter makes edges that look something like shallow finger joints (bottom detail drawing, facing page). These edges align the boards and provide a larger gluing surface than simple square edges do. Glue-joint bits are also available for use in router tables.

First I lined up the boards so that their grain matched, and I marked them so that they wouldn't get mixed up during the glue-jointing operation. I used numbers—1s on the first two adjoining edges, 2s on the next two, and so on.

I don't have a wide planer, so I had to flatten the glued-up boards with a belt sander. With a 60-grit belt, I sanded across the grain first, then with the grain. Then I used a 100-grit belt and finished with a 120-grit belt. The countertop, the most visible of these wide boards, was finished using 180-grit paper on a random-orbit sander.

Spline-and-Groove Wainscot

One of the original small rooms had beaded wainscot all the way around, so I decided to use beaded wainscot inside the hutch. To make the wainscot, I ripped ash boards on the table saw into random widths, from 5¼ in. to 3¼ in.

To join the pieces, I used a spline-and-groove joint rather than a tongue-and-groove joint (top detail drawing, facing page). First I jointed the edges of each board. To make the groove, I used a ¼-in. straight cutter on the shaper, but a ¼-in. slotting

Putting the Carcase Together

Countertop notched to accept full-length wainscot.

Wainscot divider is biscuited in place.

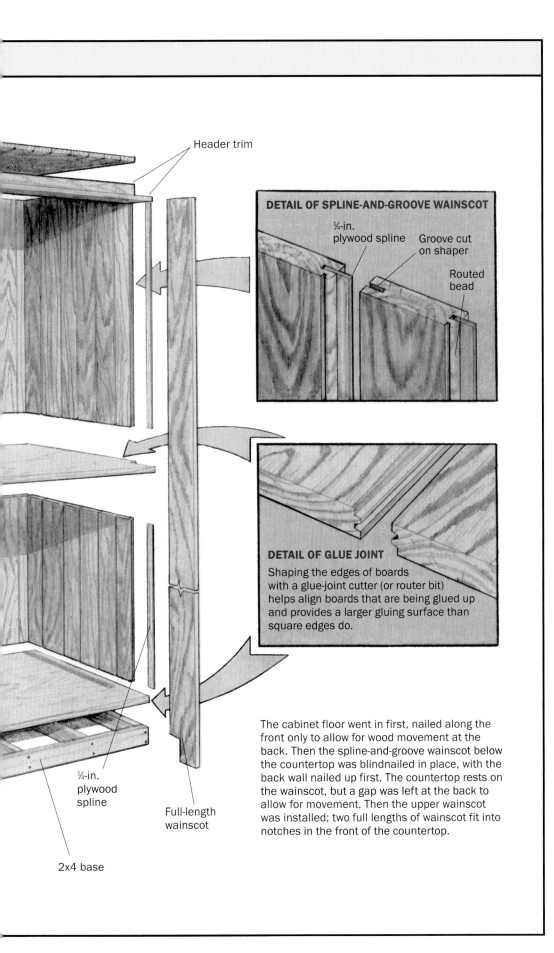

Header trim

DETAIL OF SPLINE-AND-GROOVE WAINSCOT

¼-in. plywood spline

Groove cut on shaper

Routed bead

DETAIL OF GLUE JOINT

Shaping the edges of boards with a glue-joint cutter (or router bit) helps align boards that are being glued up and provides a larger gluing surface than square edges do.

¼-in. plywood spline

Full-length wainscot

2x4 base

The cabinet floor went in first, nailed along the front only to allow for wood movement at the back. Then the spline-and-groove wainscot below the countertop was blindnailed in place, with the back wall nailed up first. The countertop rests on the wainscot, but a gap was left at the back to allow for movement. Then the upper wainscot was installed; two full lengths of wainscot fit into notches in the front of the countertop.

Trimming the Cabinet

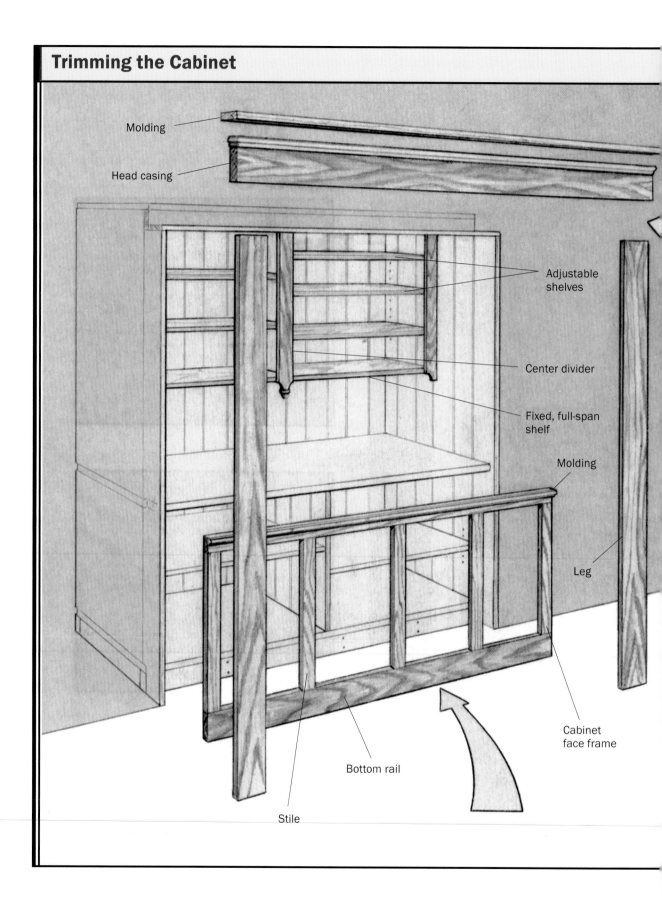

Molding

Head casing

Adjustable shelves

Center divider

Fixed, full-span shelf

Molding

Leg

Cabinet face frame

Bottom rail

Stile

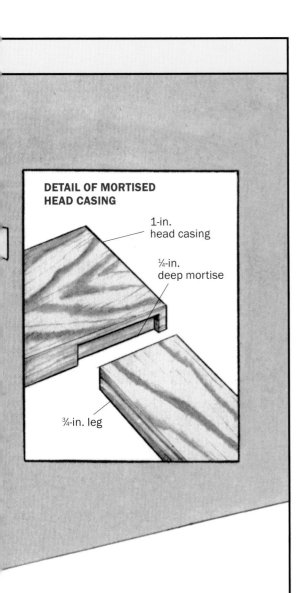

The cabinet face frame was assembled, then glued and screwed in place. The head casing was mortised to fit over the legs to keep the joints from opening up. A fixed, full-span shelf above the countertop supports a divider and adjustable upper shelves.

cutter in a hand-held router or a dado-blade assembly in the table saw would work, too. I centered the ½-in.-deep groove on the edge of the board. The 1⁵⁄₁₆-in. splines were ripped from ¼-in. plywood. I didn't use biscuit joinery because, when wainscot shrinks, gaps appear between the biscuits. A full spline looks like a solid tongue.

Using a beading bit, I beaded one edge of each board to match the original chestnut wainscot.

Installing the Wainscot

I installed the floor of the cabinet first, flush to the front of the 2×4 base. I nailed the floor at the front only and left a ⅜-in. space at the back to allow for wood expansion.

I put up the wainscot for the bottom half of the hutch by blind-nailing through the splines and into the walls as I would any T&G material. I didn't glue the splines because each piece of wainscot should expand and contract independently. This wainscot rests directly on the cabinet floor; if the floor butted into the wainscot, a seam would open. I avoided visible seams in the corners by putting up the back wall first and then butting the sidewalls into it. And I allowed for expansion by installing the first board on the back wall ⅜ in. from the corner.

I also made a wainscot divider for the bottom cabinet. It was biscuited to both the floor and the underside of the counter. I used just a dab of glue in each biscuit slot to prevent any unnecessary glue squeeze-out.

The counter sits on the wainscot. Before I installed the counter, I notched its two front edges, which would allow an entire length of wainscot at the front of each sidewall. Along the sides, the counter is nailed into the wainscot so that it stays put, but to allow for expansion and contraction, the back edge of the counter isn't nailed.

Now I was ready to put the wainscot in the top of the hutch. I set the wainscot on the counter and blind-nailed it through the splines into the walls. Putting the wainscot up in two sections, bottom and top, eliminated the wood-shrinkage gaps that would have resulted from running the wainscot from floor to ceiling and butting the counter into the wainscot. With the front edges of the counter notched, I installed the front pieces of wainscot on each sidewall. Because the unit is recessed into the opening, I wanted a full length of wainscot from floor to header with no seam.

Pocket-Screw Joinery

In my shop, stiles and rails for the face frame were cut to width but not to length. Stiles and rails are the vertical and horizontal frame pieces, respectively.

I assembled the face frames on site. I cut the stiles and the rails to length and clamped them to the cabinet to check the fit. After some slight trimming on a compound-miter saw perfected the face-frame joints, I laid the stiles and the rails on the bench and screwed them together.

The top rail was narrow enough to allow the stiles to be joined to it with screws driven straight through the edge. But the bottom rail of the cabinet was wide, and the intermediate stiles butted into it, so here I screwed the rail to the stiles through pocket holes. A pocket hole is a cut made on the face of a board that doesn't reach the board's edge.

There are several jigs on the market to make pocket holes—from simple guides for a hand-held drill to dedicated pocket-hole machines. I don't have any of them, so to make the pocket holes to assemble this face frame, I used a spade bit, starting the hole with the drill held vertically and tipping the drill back as I fed the drill bit in (top photo). The pocket hole ended at a mark 1½ in. from the edge of the rail. Then I drilled a pilot hole in the edge of the rail at an angle up through the pocket hole (middle photo). Finally, I squeezed a generous amount of glue between the stiles and the rails, clamped them together and ran the screws in (bottom photo).

After the glue was dry, I sanded the joints flush and installed the face frame. I glued the bottom rail to the front edge of the cabinet floor and screwed the top rail to the underside of the countertop (drawing, p. 84).

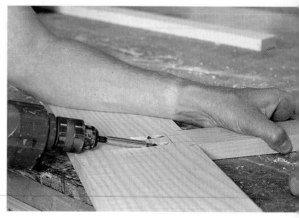

Pocket-screw joinery. To attach the bottom rail to the stiles, a spade bit makes a pocket hole that's 1½ in. short of the rail's edge (top). A pilot hole is then drilled up through the edge to connect with the pocket hole (middle), and the boards are glued and screwed together (bottom).

Mortised Head Casing

The trim, or casing, around the hutch was installed next. The ¾-in.-thick side pieces, or legs, went on first; I ran them ¼ in. long at the top. The 1-in.-thick top piece, or head, was mortised to fit over the legs (detail drawing, p. 85). You could think of this as being a mortise-and-tenon joint, with the legs being the tenons. I set the head on top of the legs and with a sharp pencil traced the outline of each leg onto the bottom edge of the head. Then I scored the marks with a sharp knife. Scoring makes for a cleaner mortise. I mortised these sections of the head a good ¼ in. deep with a hinge-mortising bit in my small router. Finally, I used a chisel to square the corners of the mortise. This joint practically guarantees lasting beauty: If the header shrinks, the joint still looks tight.

I wanted the molding under the front of the counter and at the top of the head casing to match the original molding at the top of the doors and the windows (photo, above). This molding wasn't something I could have picked up at the lumberyard, and I couldn't find any cutters the right shape, so I combined two different shaper cutters to make the molding (drawing, right). The result was a perfect match.

Making Doors

I made frame-and-panel doors for the cabinet at the bottom of the hutch.

The door stiles are 2 in., the top rail is 2½ in., and the bottom rail is 3½ in. After cutting the pieces to size, I used my shaper to mold the inside edges of the frame, cut the panel groove, and made the cope-and-stick joint between the stiles and the rails.

I assembled the frames dry to check the door size and to get the panel size. I allowed ⅛ in. on each side of the panel for expansion. The ash panels on these doors were raised (beveled around the edges) on the shaper, so I glued up the boards with square joints to make the wide panels. If I had used the glue-joint cutter, the glue-joint profile would have been visible when I shaped the raised edges.

To be sure everything fit, I dry fit the panel within the frame before gluing up. Then I glued the doors and clamped them. I used a small amount of glue on the joints because the squeeze-out could glue the panel in place, and the panel should be free to expand and contract.

Stephen Winchester is a carpenter and woodworker in Center Barnstead, New Hampshire.

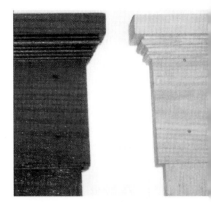

A table saw and a shaper were used to make ash molding (right) that matches the original chestnut trim (left).

Matching Molding

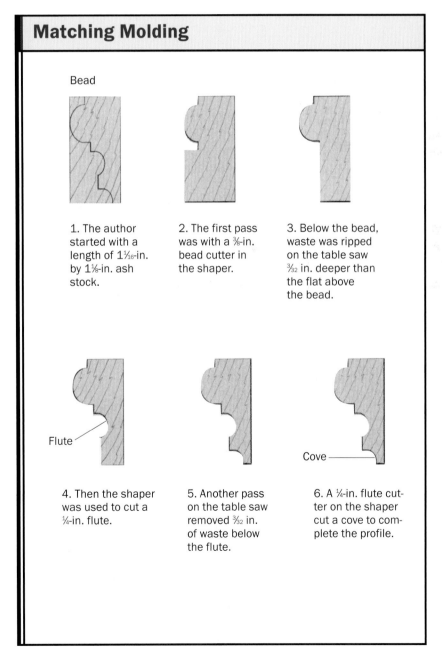

Bead

1. The author started with a length of 1¹⁄₁₆-in. by 1⅛-in. ash stock.

2. The first pass was with a ⅜-in. bead cutter in the shaper.

3. Below the bead, waste was ripped on the table saw ³⁄₃₂ in. deeper than the flat above the bead.

Flute

4. Then the shaper was used to cut a ¼-in. flute.

5. Another pass on the table saw removed ³⁄₃₂ in. of waste below the flute.

Cove

6. A ¼-in. flute cutter on the shaper cut a cove to complete the profile.

Designing and Building an Entertainment Center

■ BY BRIAN WORMINGTON

Pocket doors slide inside cabinet.

Hanging shelf provides clearance for doors.

Good looking and well organized. Video gear above the TV, audio equipment to the side, and drawers for tapes and CDs help to manage the clutter of entertainment.

Drawers for CDs and audio and video tapes.

The first extendable TV turntable I installed in an entertainment center was a surprise. This slick hardware, rated to hold 200 lb., seemed to be an engineering coup. I smiled as I helped my customer set his new 32-in. TV in place. But as he extended the turntable, I started to worry: The TV bobbed up and down alarmingly. The bobbing worried my customer, too. He never again pulled the TV out from the cabinet—the image of his $900 TV on the end of a diving board was too disturbing.

I run a one-man shop where custom entertainment centers are my main business. Years of handling the whole process—designing, building, and installing—have taught me to avoid such gaffes as flexible turntables.

Most Entertainment Centers Revolve Around the TV

Designing an entertainment center begins with a visit to the client's home. I measure the space, see where the TV will be most easily viewed and verify the presence of electrical and cable outlets. Sometimes there is a radiator or heat vent in the way, and the client needs to be reminded to call the appropriate contractor to move it.

I build freestanding and built-in units. My techniques apply equally to both types. Because I work alone, I build entertainment centers in modules (drawing, p. 90). One large cabinet would be difficult to handle, but I can move smaller modules by myself and join them together on the job.

The design of most entertainment centers revolves around housing a 27-in. or 32-in. TV (drawing, above). Such large TVs are 2 ft. or more deep. Although the dimensions don't vary much among TV manufacturers, I always measure the set before designing the cabinet.

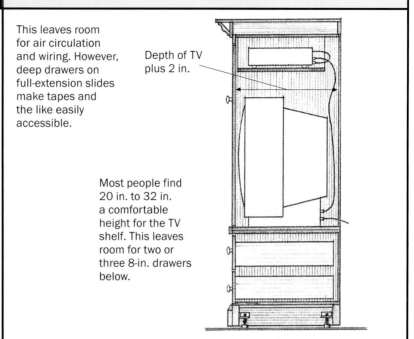

Component Shelves Don't Go the Full Depth of the Cabinet

This leaves room for air circulation and wiring. However, deep drawers on full-extension slides make tapes and the like easily accessible.

Depth of TV plus 2 in.

Most people find 20 in. to 32 in. a comfortable height for the TV shelf. This leaves room for two or three 8-in. drawers below.

Depending on the final height of the TV shelf, I fill the space below it with two or three full-extension drawers to store CDs, audio tapes, videocassettes, and accessories.

CDs are 4⅞ in. wide, and audio tapes are 4¼-in. wide. I divide my standard 20¼-in.-deep drawers into 5-in. rows that accommodate either. Videotapes take up 7½ in., so I divide drawers for them into three rows. The front two rows are the full width of the tapes, while the back row holds several rows of four tapes lengthwise. Reversible blocks inside the drawers hold aluminum bars that divide tapes and CDs when set one way, and videocassettes when set the other way (photo, below).

Aluminum bars divide drawers. These drawer dividers are set up for videotapes, but the wood blocks where the bars rest can be flipped over. Their other side has slots to arrange the bars to organize CDs and audio cassettes.

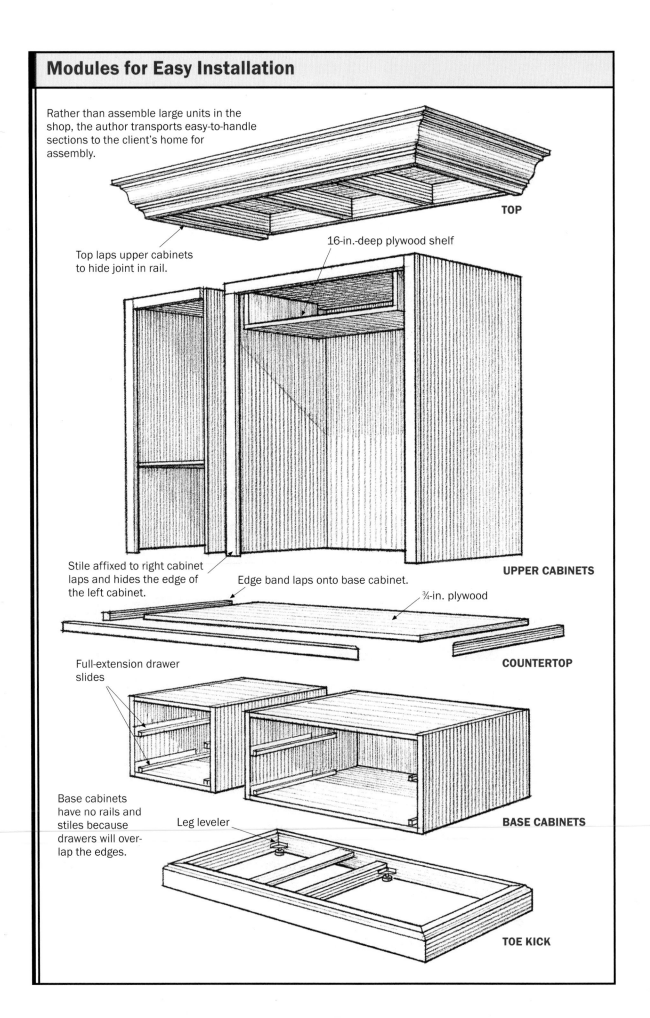

Modules for Easy Installation

Rather than assemble large units in the shop, the author transports easy-to-handle sections to the client's home for assembly.

TOP

Top laps upper cabinets to hide joint in rail.

16-in.-deep plywood shelf

UPPER CABINETS

Stile affixed to right cabinet laps and hides the edge of the left cabinet.

Edge band laps onto base cabinet.

¾-in. plywood

COUNTERTOP

Full-extension drawer slides

Base cabinets have no rails and stiles because drawers will over-lap the edges.

Leg leveler

BASE CABINETS

TOE KICK

Hidden levelers ease installation. After assembling the entertainment center in the house, the author adjusts the levelers through holes that will be concealed by drawers.

Biscuits and Pocket Screws Hold the Cabinet Together

I build these units from ¾-in. hardwood-veneer plywood. I butt-join the tops, bottoms, and sides, using #20 biscuits to align the parts and pocket screws to hold them together. To keep the case square, I rabbet the sides, top, and bottom to receive a ½-in. plywood back. I screw the back to the case with 1⅝-in. cabinet screws. The back is strong enough for the 2½-in. screws I use to attach the cabinet to the wall. I don't need an extra cleat. If necessary, the back can be removed easily to route electrical wires.

Put the VCR and the TV in the Same Module

Once the location of the TV is decided, the arrangement of the other equipment more or less falls into place. I find that the best place for the VCR and the cable or satellite box is on a shelf above the TV. This placement simplifies wiring, and the TV and its peripherals can hide behind the same doors. That way, if the TV is on, the doors in front of the VCR are open. This is important

because remote controls communicate with their components by line-of-sight signals. There are devices that sidestep this problem by relaying the signal. But they cost about $100, and the need for them is avoided by keeping the VCR with the TV.

If there are only a few components, it may be possible to stack the audio and the video gear together on a shelf about the TV. Most equipment is the same width, 17¼ in., and about 1 ft. deep, so it stacks well.

But few people trouble with an entertainment center to house only a few components. They either own or intend to buy a serious system. I often build three-door cabinets to house such systems. Two doors enclose a TV cabinet, and the remaining door closes on an audio-component cabinet about half the width of the TV cabinet.

Books Are Entertainment, Too

Often, clients want bookshelves as part of their entertainment centers. Bookshelves generally aren't as deep as shelves for audio or video components; a 12-in. shelf accommodates most books. I build plywood bookshelves in modular units no more than 32 in. wide, a dimension that minimizes plywood waste. It's also a maximum width that I'm comfortable with for a ¾-in. plywood shelf. Wider shelves sag more under the weight of books. Of course, it's possible to build wider shelves that don't sag by laminating several layers of plywood together. Another way to beef up a shelf is to replace the regular ½-in. by ¾-in. hardwood edge with 1×2 or 1×3 stock.

Keeping the Music Cool

A frequent concern my clients have is cooling their electronic equipment and TV. This concern is a holdover from the days of tube-based equipment, which generated considerable heat. Modern solid-state components

Sources

Accuride International
12311 Shoemaker Ave.
Santa Fe Springs, CA 90670
(562) 903-0200
www.accuride.com

Julius Blum & Co.
P.O. Box 816
Carlstadt, NJ 07072
(800) 526-6293
www.juliusblum.com

JDR Microdevices
1850 S. 10th St.
San Jose, CA 95112
(800) 538-5000
www.jdr.com

Knape & Vogt Manufacturing Co.
2700 Oak
Industrial Dr. NE
Grand Rapids, MI 49505
(616) 459-3311
www.kv.com
KV 1385

Titus
22020 72nd Ave. S.
Kent, WA 98032
(800) 762-1616
www.titusint.com

don't generate much heat. Some manufacturers recommend about 7 in. of airspace above the components and a few inches to the sides and back.

Sometimes, however, I encounter tube-based equipment and large power amplifiers. They need air movement; the manufacturer can tell you how much. In these cases, I drill ventilation holes in the top or in the back of the cabinet and ventilate with whisper fans. These small fans are commonly used to cool computers. I buy them for about $12 each from JDR Microdevices®.

TVs don't need much ventilation. The picture tube is taller than the vents in the TV's back, so there is always airspace above. And the cabinet doors are open when the TV is in use, so airflow is constant.

Provide Access to Equipment Backs

I encourage my clients to stack their components—except for heat-generating tube equipment—atop each other. I provide a pull-out shelf on a full-extension drawer slide to stack the components on. With the shelf extended, there is easy access to rear-panel cable connections.

An exception is when the client has a record turntable or carousel-type CD player that loads from the top. I provide individual pull-out shelves for these items.

Hiding the TV

Many of my clients want to be able to close the TV behind doors. I think it's to hide the fact that they actually watch TV. Sometimes, regular hinged doors won't work. There may not be room to open them all the way, or the client simply may not like the look of open doors. This is why pocket doors that open 90° and slide into the cabinet alongside the TV are a popular option (bottom left photo). The cabinet width must be increased by 2 in. on a side, or 4 in. total, to accommodate the pocket-door hardware.

Because pocket doors slide into the cabinet, they are necessarily smaller than their opening. Therefore, they can't overlay the cabinet face and must be inset. For symmetry, I often inset the other doors on the unit as well, using Blum® European-style hinges made for inset doors.

There are several types of pocket-door hardware on the market. The most common and least expensive resembles paired drawer slides. I use them only for doors up to about 30 in. high. Larger doors sag when extended and rub on the cabinet bottoms. Accuride and Blum both make versions of this slide.

For larger doors, I use Accuride model 1332 hardware. The 1332 uses a set of cables like those on a draftsman's parallel rule to keep the hinges perfectly aligned.

Accuride suggests a ⅛-in. margin between the doors and the cabinet, but I think 1/16 in. looks better. In entertainment centers with

Pocket doors hide that embarrassing TV when guests visit. Then they slide unobtrusively to the sides of the TV as the set returns to regular use.

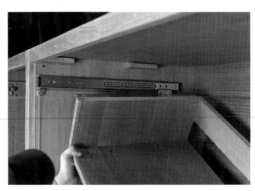

Shelf hangs from cabinet top. Angle iron on the shelf's side fits into rabbeted blocks above the pocket-door slides.

Shelf doubles as a doorstep. Stops on the countertop might interfere with sliding in the TV.

pocket doors, I hang a shelf unit from the top of the cabinet (right photo, facing page). This provides a pull-out shelf above the TV for the VCR while leaving space for the pocket doors to slide.

Making the Best of Extendable Turntables

The story at the beginning of this article doesn't always discourage my clients from turntables. If they insist, I buy the heaviest-duty turntable I can find, usually the KV® 1385. The cost difference between this 200-lb. rated turntable and a 150-lb. rated turntable is less than $5. I fasten turntables to the cabinet with ¼-in. #20 bolts and washers. Wood screws will eventually strip out, dropping that $900 TV to the floor.

A more stable platform for extending the TV from the cabinet is a lazy Susan affixed to a shelf. The shelf is mounted on 200-lb. rated full-extension drawer slides. This set-up is much steadier, but it can't be used on entertainment centers that have pocket doors. The drawer slides must be screwed to the sides of the cabinet and would leave no room for the pocket doors.

Brian Wormington owns Acorn Woodworks in Otis, Massachusetts.

Adjustable Shelves That Stabilize a Tall Bookcase

To accommodate clients who are interested in showing off extensive libraries, I build bookcases. To stiffen wide or tall bookcases, it's sometimes necessary to fix a centrally located shelf in place. Rather than permanently install such shelves, I preserve some adjustability by affixing the shelf with Titus knock-down connectors. These connectors have studs that thread into shelf-peg holes. Connectors that are set into holes bored in the shelf bottom capture the studs. A reverse turn of the screwdriver drives the parts tightly together, bracing the cabinet and making it stable for plenty of books.

Case sides are bored for shelf pegs on 32mm centers. The studs for these connectors screw into the peg holes. They can be moved to different holes to improve the shelf layout, something that's impossible with fixed shelves.

Building a Fireplace Room Divider

■ BY ALEXANDER BRENNEN

Josephine Coatsworth had an enviable problem. Most people who expand their kitchens have to make hard choices about how to gain the extra space. Do I take over the laundry room or build an addition? Josephine's dark, narrow kitchen, however, was next to an 18-ft. by 34-ft. spare room. Rumor had it that the prior owners of the house had used this empty space as a dog run. A partition wall with a small door separated the two rooms. Because there were no openings or windows in the partition, you wouldn't know from this gloomy corner of the kitchen that there were three sliding glass doors in this spacious room that opened to the backyard.

Josephine envisioned the unused space as a family room. Even with 6 ft. annexed from one end for enlarging the kitchen, the remaining space would be 28 ft. long—big enough for a family room with a ping-pong table and a corner for the television. For chilly evenings, Josephine wanted a fire-place in the family room situated so that it could be seen from the kitchen.

With those requirements in mind, architect Bennett Christopherson devised a plan that divides the two rooms with a two-sided fireplace. Built-in benches extend from the fireplace on the kitchen side of the partition, forming an L-shaped seating area that wraps around a breakfast table (top photo, facing page). The space above the bench is open to the family room, which lets light into the kitchen and mingles the activities of both rooms.

The backs of the benches become a large shelf. On the family-room side of the partition, a cabinet for the television tucks under this broad shelf, and a niche under the fireplace serves as firewood storage (bottom photo, facing page). The entry to the family room is adjacent to the fireplace through a deep passageway that is defined by a low-ered soffit. A small closet facing the family room completes the divider.

A multipurpose partition. Built-in benches topped with a wide display shelf define the corner of the remodeled kitchen. Because it's two-sided, the fireplace can be enjoyed simultaneously from the kitchen and from the family room.

On the family-room side. The fireplace sits atop a raised hearth, which has been left open below for firewood storage. The carved doors to the left of the fireplace conceal the television and the stereo equipment and provide space for tape and CD storage.

The Frame Under the Plaster

Once the overall concept was on paper, my first job was to assess the structural role of the original wall between the kitchen and the empty room. I was relieved to learn that it was merely a partition, with no load-carrying or earthquake-bracing duties. Instead of carrying the load of the roof, the partition had to be strong enough to support a television, a prefabricated fireplace, and people sitting at the breakfast bench. I yanked out the old partition and considered the starting point for the new framing.

The bench height is the most important horizontal line in the partition's overall composition. So before I started assembling the frame, I mocked up some sample bench

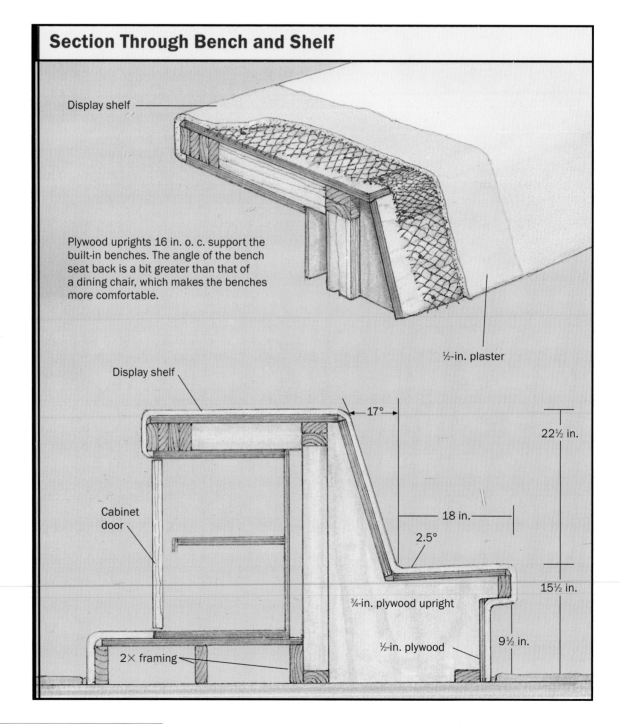

Section Through Bench and Shelf

Display shelf

Plywood uprights 16 in. o. c. support the built-in benches. The angle of the bench seat back is a bit greater than that of a dining chair, which makes the benches more comfortable.

½-in. plaster

Display shelf

Cabinet door

17°

22½ in.

18 in.

2.5°

15½ in.

¾-in. plywood upright

½-in. plywood

9½ in.

2× framing

profiles to find the right height and seat-back angle. Because the benches are used for relaxing as well as seating for meals, I made the seat backs tilt back a little farther than a standard dining chair would.

As I tested the various mockups, I placed 3-in. cushions on them to duplicate the thickness of the cushions on the finished benches. I eventually settled on a bench height of 15½ in. (drawing, facing page). This line extends beyond the benches to become the fireplace hearth.

There's nothing mysterious about the partition framing. It is made out of standard 2× material—mostly 2×4s on 16-in. centers for both horizontal and vertical members (top photo). The bench is supported by ¾-in. plywood uprights on 16-in. centers.

With its stucco walls and courtyard plan, Josephine's house feels a little like a hacienda. To be consistent with that style, she wanted the partition to have rounded corners and to be finished with hand-tooled plaster. That meant the framing members had to be radiused at the corners. Most corners have a 3-in. radius. The bench, on the other hand, has a 6-in. radius on its one open corner, as does the shelf on the family-room side (bottom photo).

A Prefabricated Fireplace

Prefabricated steel fireplaces don't weigh nearly as much as masonry fireplaces. Consequently, they don't require the heavy footings demanded by traditional fireplaces. I checked with the supplier of the fireplace to find out what kind of support structure was needed and what clearances needed to be considered. For this fireplace insert, 2×4s on edge were enough to support the 50-in. span of the insert.

I sheathed the fireplace platform and its 20-in.-deep hearths on both sides of the firebox with 3¾-in. CDX plywood. Next I covered this plywood with 26-ga. galvanized sheet metal, followed by a layer of cement

No masonry needed. One of the advantages of a prefabricated fireplace is that it can be supported with simple stick framing. The height of this hearth was determined by the height of the bench.

Southwestern means curves. In order for the plaster to curve properly around the edges of the wood framing, the plywood sheathing has to be held back a bit from the framing members.

Sources

Marco Fireplace Distributing
13018 Rayner St.
N. Hollywood, CA 91605
(818) 764-3366
www.marco-fireplace.com
Marco delux see-thru

fiberboard. These layers of noncombustible materials satisfy the local code requirement for nonflammable surfaces under, and in front of, a woodstove or a steel fireplace insert.

Side clearances for the two distinct portions of the fireplace—the firebox and the flue—are also important. With this fireplace, I had to make sure there were 2 in. between the flue and any combustible materials and 1 in. between the sides of the firebox. The nailing flanges of the fireplace, on the other hand, can be nailed directly to the framing. To get the required clearances, I had to break out the reciprocating saws and cut chamfers and curves on some of the framing members.

When he installed the fireplace, the subcontractor also ran the double-wall stainless-steel flue through the roof, where it is topped with a rainproof chimney cap. At the same time, he ran a 4-in.-dia. aluminum flex pipe to the soffit, where it takes in combustion air for the fireplace (photo, right). Then the plumber took over and installed a gas log-lighter.

I think prefabricated steel fireplaces are pretty good values. This one lists for about $1,400, and it's a deluxe model. An economy-grade, non-see-through fireplace from Marco can cost as little as $500. That's a big difference in price from a masonry fireplace, especially when you consider the cost of the footings. But don't forget the stainless-steel flue. It costs in the range of $70 per 4-ft. section. Flashing kits are about $50, and a chimney cap costs around $60.

Finished with Plaster

The plasterers used galvanized expanded metal lath and stucco wire to form an armature for the three coats of plaster that cover the partition (drawing, p. 96). The stucco wire covers the large planes, and the expanded metal lath reinforces the corners and the edges where the plaster dies into cabinet doors.

The small duct delivers outside air. A double-wall stainless-steel flue runs cool enough to allow the wood framing to come within 2 in. of the flue. The smaller duct delivers combustion air to the fireplace. Note how the hearth framing has been changed from that shown in the top photo, p. 97, to accommodate the television.

The drywall texture in the rest of the room was made to match that of the partition by applying joint compound in sweeping strokes with a large trowel. Once dry, the trowel marks were sanded down a bit and then recovered with thinned topping mud applied with a trowel. This produced a pleasing, undulating plastered look that is in keeping with the hand-tooled surface of the partition. Now that they're covered with a couple of coats of latex paint, the textures of the walls, the ceiling, and the partition flow easily together.

Cost estimates are from 1994.

Alexander Brennen is a general partner of Zanderbuilt, in Berkeley, California.

Improving Kitchen-Cabinet Storage

■ BY SVEN HANSON

The shelves in kitchen cabinets break a fundamental design rule (at least the lower cabinets, anyway): A storage space shouldn't be much deeper than the height of the space above it. Headroom in a base cabinet with one shelf is typically less than 11 in. but twice as deep. The result is pans hidden behind pots, a colander that hasn't been seen since 1999, and sore knees.

Over the 25 years that I've been a cabinetmaker, I've devised many ways to resolve the cabinet conundrum. You can improve base cabinets tremendously by adding roll-out shelves, but why stop there?

Base-cabinet drawers can be improved as well. Full-extension hardware maximizes utility, and dedicated inserts can send the convenience of drawers off the charts. A spice rack, a pot-lid drawer, a roll-out wastebasket, and even a pullout undercabinet pantry are all simple to build. Dovetail joinery isn't necessary. Polyurethane glue, screws, and nails make strong joints, and they'll look fine if you're the slightest bit conscientious about hiding fasteners.

Easy-Access Trash Can

The undersink space is a classic coal pit wanting to be a gold mine. There's just enough room to squeeze a trash can in front of the drainpipes and to pack cleaning supplies under them. A roll-out drawer can improve this space dramatically. A divider steadies the trash can and creates the rear storage compartment. The tricky part is mounting the drawer hardware. By making a cradle, you can sidestep expensive bottom-mount hardware. Because the cook will be throwing garbage at the trash can, you should use heavily finished wood and shouldn't glue the cradle to the cabinet to ease disassembly and cleaning.

Door

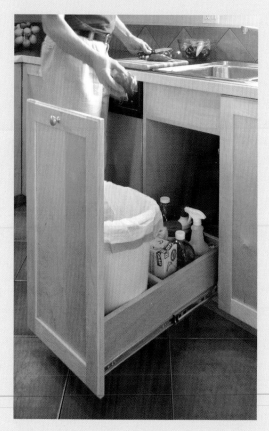

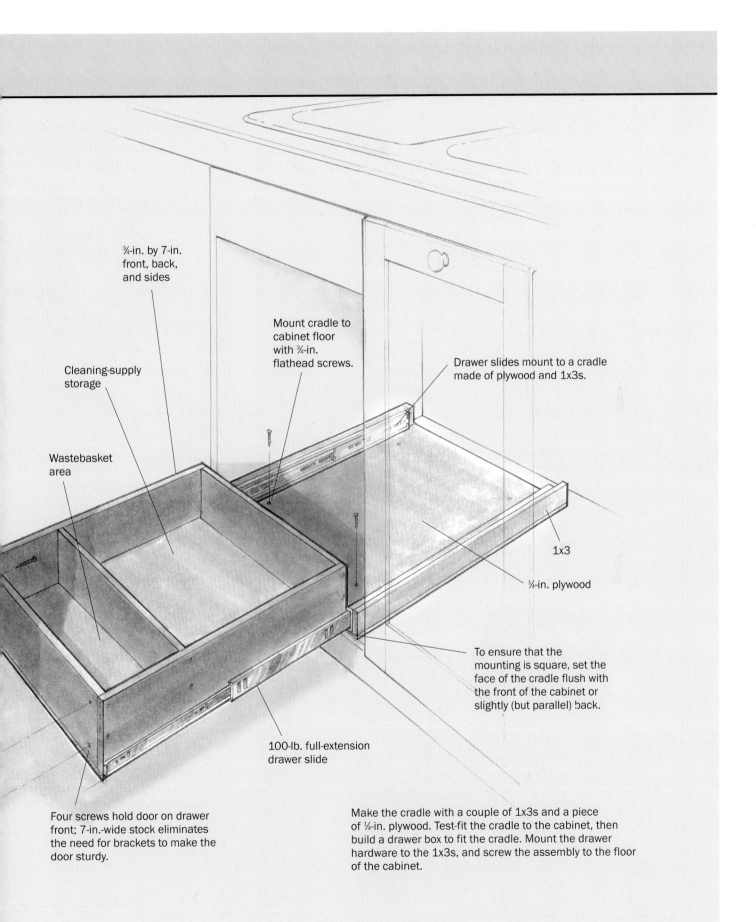

¾-in. by 7-in. front, back, and sides

Cleaning-supply storage

Wastebasket area

Mount cradle to cabinet floor with ¾-in. flathead screws.

Drawer slides mount to a cradle made of plywood and 1x3s.

1x3

¼-in. plywood

To ensure that the mounting is square, set the face of the cradle flush with the front of the cabinet or slightly (but parallel) back.

100-lb. full-extension drawer slide

Four screws hold door on drawer front; 7-in.-wide stock eliminates the need for brackets to make the door sturdy.

Make the cradle with a couple of 1x3s and a piece of ¼-in. plywood. Test-fit the cradle to the cabinet, then build a drawer box to fit the cradle. Mount the drawer hardware to the 1x3s, and screw the assembly to the floor of the cabinet.

Inserts Organize Spices and Spoons

I've built drawer inserts many ways, and I like this method best because it scales up or down, is convenient for a broad array of drawer sizes, can be removed to clean, looks good, and works well.

The key concept is that it should be too small for the drawer. A small insert slips in easily and creates an additional silverware slot. Attach the insert to the front and one side of the drawer.

Build it with ⅜-in. stock, 2 in. wide. Three-in.-wide partitions make it easy to reach, even for people with big hands. Assemble in any pattern using two #4 screws at each joint.

A Sporty Spice Rack

Spices should be stored in a cool, dark place, not on top of the back of the stove, where my editor stores his. But they also need to be close at hand for cooking. A small drawer located next to the stove is the perfect place, and angled shelves make an excellent storage system.

With scraps left over from other projects, you can build a rack that lets the spices gently recline until called to service. Use ¾-in. screws for hidden fasteners and finish nails for exposed spots. I build it to go all the way across the drawer, but you can cut it short and add a ¾-in. side along the cut edge, opening the rest of the drawer for other uses.

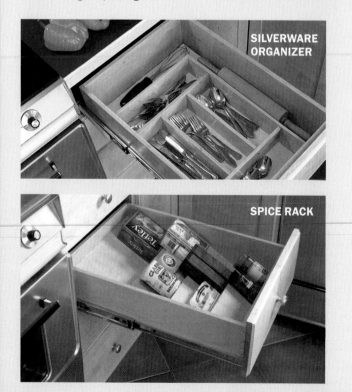

SILVERWARE ORGANIZER

SPICE RACK

The Roll-Out Pantry

Three simple drawer boxes, an 8-ft. piece of aluminum angle, and some heavy-duty full-extension drawer slides are all you need to convert a barely useful 12-in. base cabinet into one of pure utility. Cut the aluminum into four equal pieces, and predrill holes for flathead wood screws. Fasten the aluminum angle to each corner of the bottom drawer, starting 2 in. above the bottom of the box to steer clear of the hardware. The middle shelf drops in from above and is easily set with a couple of 7½-in. spacer blocks. The top box should have tall sides to resist racking and to provide added protection for glass items on top.

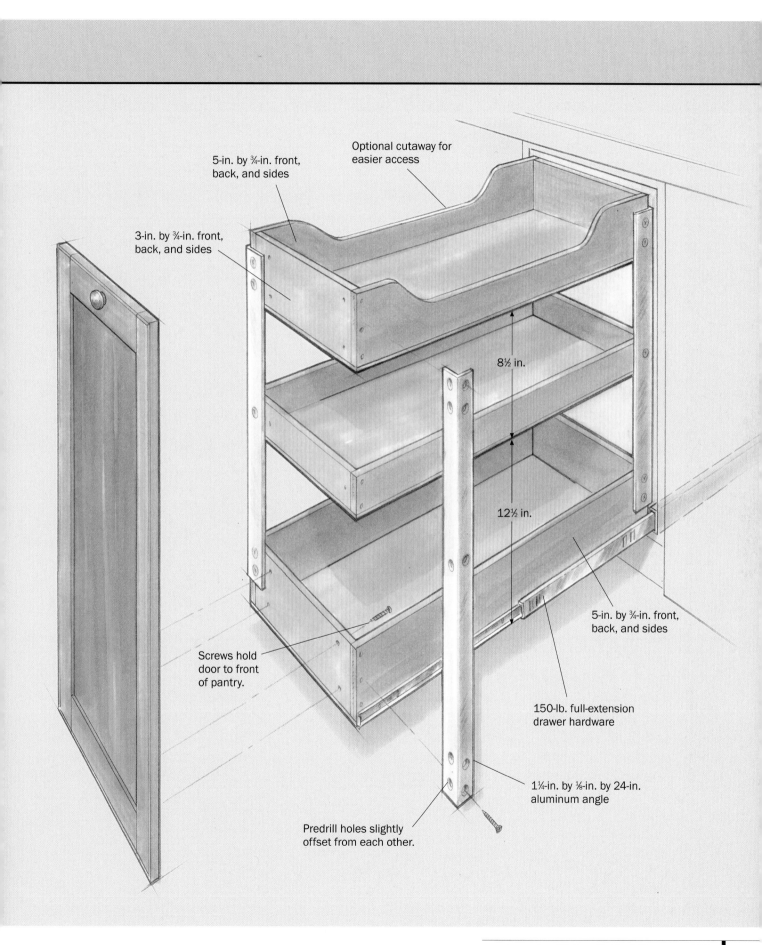

5-in. by ¾-in. front, back, and sides

Optional cutaway for easier access

3-in. by ¾-in. front, back, and sides

8½ in.

12½ in.

5-in. by ¾-in. front, back, and sides

150-lb. full-extension drawer hardware

Screws hold door to front of pantry.

1¼-in. by ⅛-in. by 24-in. aluminum angle

Predrill holes slightly offset from each other.

A Drawer for Pots and Lids

This insert can slip into existing drawers. Uneven louver spacing gives more flexibility. To build one, I rip enough ¾-in. stock for three 5-in.-wide boards to fit the drawer front to back. After sanding the prettiest three sides, I rip the ugly edge at 40° and screw two strips of ¼-in. plywood to the beveled edges (no glue allows adjustment).

Louvers are 5 in. tall and long enough to slip into the drawer.

EFFICIENT DRAWER BOXES

Everyone has their own way of building a drawer box, but here's my take. Some people spend hours making furniture-grade dovetail joints. But where the point of the project is strength, durability, and affordability, I simplify the construction. Polyurethane glue and screws are remarkably strong, simple, and slow-drying—perfect for the insecure, slow carpenter.

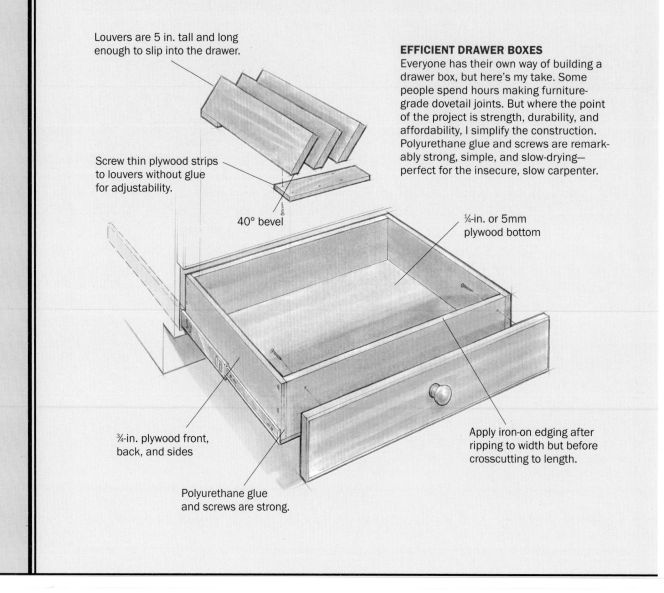

Screw thin plywood strips to louvers without glue for adjustability.

40° bevel

¼-in. or 5mm plywood bottom

¾-in. plywood front, back, and sides

Polyurethane glue and screws are strong.

Apply iron-on edging after ripping to width but before crosscutting to length.

You Can Single-Handedly Grab the Pot Lid You Need

In use, my pot-lid holder looks like giant louvers with metal and glass Frisbees® stuck in them (photo, p. 99 and drawings, p. 104). The louvers are set perpendicular to the drawer front and can slide into an existing drawer or into a newly assembled one. The drawer side becomes the final louver once you drop the assembly into place. This adjustable design (which uses no glue, only screws) allows three rows of lids and space alongside for pots and pans. The louver assembly can be removed for easy cleaning.

Roll-Out Wastebasket Drawer

Undersink space is fertile territory for improvement. The disposal, drainpipes, and trap conspire to destroy almost all hope of organized storage. But you can help these lost cubic feet achieve their full potential with a bottom-mounted roll-out wastebasket drawer. A wastebasket drawer makes use of the tall space in front of the plumbing and the low space under the sink trap to the rear of the cabinet box (photo, p. 100).

The challenge with this retrofit is mounting the hardware. Sink bases are typically wide, frequently with a center divider. One solution is expensive drawer hardware that mounts to the floor of the cabinet. My solution is a self-contained unit made of plywood and 1x3s, which holds the standard drawer hardware and drops into the base cabinet after assembly (drawing, p. 101).

To keep it simple, I screw the cabinet door to the face of the wastebasket drawer. Let the door hang ⅛ in. below the top of the toe kick. You can now pull it out with your foot when you're elbow-deep in juicy tomatoes.

Roll-Out Pantry: a Home for the Cheerios®

Skinny base cabinets are perhaps the most wasteful use of space. Not only are they dark and deep, but they're also too skinny to see into. Yes, you can use them for cookie sheets and cutting boards, but I think there's a better use: a roll-out pantry. With access from both sides, a narrow unit prevents stored goods from being buried: They're all easy to reach (right photo, p.102).

After determining the width and the depth available, build three drawers, two at 5 in. tall and one at 3 in. tall. The drawers are fastened vertically to each other with aluminum angle at each corner to form a single box that has three trays.

Drilled with the hole pattern as shown in the drawing on p. 103, the aluminum angle becomes the shelf standards, and it attaches to both the front and the sides of the drawers.

Sven Hanson is a contributor to Fine Homebuilding *and* Fine Woodworking. *He's a cabinetmaker in Santa Fe, New Mexico, and Atlanta, Georgia.*

Updating the Kitchen Pantry

Every kitchen needs a pantry. A floor-to-ceiling cabinet with either shop-made roll-outs, like this one, or manufactured components offers a great deal of convenient food storage without sacrificing much floor space in the kitchen.

■ BY DAVID GETTS

It's hard to forget the first big kitchen pantry I encountered 20 years ago. This one was a marvel of interlocking wooden components—swiveling shelves, pullout storage bins, spice racks—all in a dark-stained, varnished wood. It represented a daunting amount of labor, which probably explains why I don't build many pantry interiors like that these days.

Pantries are among the most useful cabinets in any kitchen; the floor-to-ceiling repositories hold everything from canned soup to tubs of rice. They can consist entirely of roll-out shelves (photo, left) or include a combination of shopmade shelves, roll-outs, and manufactured components such as roll-out wire shelving and baskets, door-mounted racks, and spice organizers. Available from at least three major manufacturers (see sources of pantry hardware, p. 110), these components fit in either traditional face-frame or European-style frameless cabinets.

Pantry components made for the cabinet industry offer many food-storage options. There's no reason that small shops should not take advantage of what's there. All you really need is a box to mount the components in—that's what makes them such an attractive alternative to building your own.

Wire Baskets Are Lightweight and Simple to Install

Epoxy-coated baskets (photo, right) are made with integral drawer slides that are screwed to the inside walls of the cabinet. Baskets can be mounted individually in a cabinet rather than as a set, so you can put them wherever they work best. And because they are made of wire mesh, they give the illusion of more space inside the cabinet. The downside, of course, is that a basket will not contain spills (a consideration if you have young children who may not always put away the Cheerios right side up). Wire baskets come in a variety of widths. They are one of the least expensive options among manufactured pantry components.

Some shopmade components still make sense, particularly if you build them from the same materials that are used throughout the rest of the kitchen. Adjustable wooden shelves and roll-outs are not difficult to make, and they help to blend the pantry with the other kitchen cabinets. Interestingly, though, many of the people I make cabinets for prefer wire components over wood—they find them easier to clean.

You Need to Start with a Well-Designed Cabinet Box

One of the big advantages of building custom cabinets is that you can build exactly what you want. But remember that manufactured pantry components are intended to fit inside cabinets in one of several standard outside widths: 24 in., 30 in., 36 in., 42 in., and 48 in. Fitting components into a really

Roll-Out Units Often Fit Several Cabinet Heights

These units are similar to the baskets except that all the drawers are installed as a set. They are a good choice when you want to fill a good bit of a cabinet interior with a single component. These units, such as the Rev-a-Shelf® Pull Out Pantry, are easy to install, and they work well as a retrofit for an existing cabinet because the frame telescopes to fit a range of interior cabinet heights. These shelves take up less space than shop-fabricated drawers. Installation is speedy because you don't have to measure the location for each drawer; just screw in the frame and put the shelves in their indexed slots.

odd-size cabinet may require a lot of fussing with shims and spacers to get hardware to fit. It's smart to have the hardware on hand as you design and build the pantry.

Pantries are usually full-height units that run from the floor to the top of the upper cabinets. I try to design the cabinet so that the main doors are no taller than 5 ft. Doors more than 60 in. high have a greater tendency to warp, and accessibility is reduced for anything too far off the floor.

When a pantry is integrated with standard kitchen base units, it usually is 24 in. deep so that all the cabinet faces line up. One advantage with the 24-in. depth is the budget. Both sides of the cabinet can be cut from a single sheet of plywood or melamine. But when I have a choice, I like to make pantry cabinets 26 in. deep. Here's why: With a standard 1½-in. overhang on a 24-in.-deep cabinet, the countertop extends 25½ in. from the wall. By making the pantry 26 in. deep, the edge of the counter can die into the side of the cabinet rather than project out from the cabinet edge.

Whether you choose plywood or melamine for the cabinet, you should use ¾-in. stock. Melamine, a low-pressure laminate bonded to a particleboard core, is cheaper than plywood and easy to clean. The downside is that it chips easily and can be damaged when exposed to standing water. Plus, it's heavy. The alternative is hardwood plywood, which is lighter and easier to handle, is stronger, and cuts cleaner. If I use plywood, I prefer maple because of its light color and subtle figure. Stay away from open-grained woods like oak. They are more difficult to clean.

Swing-Out Units Require a Center Partition

Larger and more elaborate, swing-out racks such as the Real Solutions Pantry Mate require a vertical panel in the center of the cabinet. Racks are hinged to this panel, and they swing completely out of the cabinet, not only providing easy access to anything you store on them but also allowing you to install narrow shelves behind the racks at the back of the cabinet. Swing-out racks are ideal for storing things you want to see easily, but their lighter-weight construction makes them best suited to smaller items.

Sources

Real Solutions
2700 Oak
Industrial Dr. NE
Grand Rapids, MI
49505
(616) 459-3311
www.kv.com

**Hafele America
Company**
3901 Cheyenne Dr.
P. O. Box 4000
Archdale, NC 27263
(800) 423-3531
www.hafeleonline.com

Rev-A-Shelf
P. O. Box 99585
Jeffersontown, KY
40299
(800) 626-1126
www.rev-a-shelf.com

Julius Blum & Co.
P. O. Box 816
Carlstadt, NJ 07072
(800) 526-6293
www.juliusblum.com

**Accuride
International**
12311 Shoemaker Ave.
Santa Fe Springs, CA
90670
(562) 903-0200
www.accuride.com

Shopmade Shelves Are a Basic Pantry Option

Shelves make the most of storage space, although getting to the back of a 24-in.-deep shelf isn't always easy. Shelf material should be a minimum of ¾ in. thick—anything less will sag. A ¾-in. shelf should never span more than 36 in. if unsupported in the middle. If a shelf will be wider than that, either add solid-wood edging to the face or use 1-in.-thick material. Veneer-core plywood resists sagging better than particleboard or medium-density fiberboard.

Shelves made from plywood or melamine should be edgebanded. For plywood cabinets, I edgeband either with wood veneer of the same species or ¼-in.-thick solid wood (the ¼-in. material is more durable). If you're using melamine, the banding material can be a matching plastic laminate, PVC, or melamine edgebanding.

Adjustable shelves make the cabinet more flexible, and they simplify construction. The two basic approaches are to plow a dado to receive a metal or plastic shelf standard, or to drill evenly spaced holes for shelf pins. I prefer the shelf pins because they're less expensive and because they look better.

Plastic spacers are neat and uniform. These ¾-in. spacers move the side of the drawer in a little so that it won't clip the edge of the open door.

Drawers Make Access to the Inside of a Pantry Easy

Roll-outs are shallow drawers concealed behind doors, and they make it much easier to get to stuff stored at the back of a cabinet. Although you can buy manufactured units, you may want to build your own so that they match materials used in the rest of the kitchen. Drawer height is your call, but 3 in. is a good standard. When using three-quarter extension slides in a frameless cabinet, the width of the drawer is typically 2 in. less than the cabinet interior. That allows

When space is tight, get the right hinge. A zero-protrusion hinge from Blum allows a cabinet door to swing completely out of the way for a roll-out shelf.

1 in. for the drawer slides (½ in. on each side) and a ½-in. spacer on each side so that the drawer doesn't bang into the door as it's opened (top photo, facing page). For full-extension slides, I like to use a heavier-duty ¾-in. spacer, which requires that the drawer box be 2½ in. less than the cabinet opening. Spacers are available from Julius Blum & Co.

You also can use zero-protrusion cup hinges from Blum, which allow the door to swing completely out of the way (bottom photo, facing page). In that case, drawer slides can be mounted directly to the interior of the case—no need for any spacers.

A three-quarter extension slide is mounted to the bottom of the drawer side, so it hides the bottom edge of the drawer. You can simply glue and staple the ¼-in. plywood bottom right to the drawer box. These drawers are good only for light things. I think pantry roll-outs should have a ½-in. bottom. Full-extension slides mount to the side of the box, not the bottom edge, requiring that the bottom be let into a groove around the inside of the drawer box.

The three-quarter extension slides are less expensive. I've had good luck with the Blum #210, which is a white, epoxy-coated slide. They operate smoothly and are easy to install. But they allow easy access to only three-quarters of the drawer box. I encourage people to spend a little more money and get full-extension slides. I use model #3834 from Accuride. Full-extension slides operate smoothly on ball bearings and have a higher weight rating than three-quarter extension slides (typically 100 lb. vs. 75 lb.).

David Getts is an author and cabinetmaker living in Bothell, Washington.

Door Ladders Make the Most of Interior Space

Why waste all that space on the back of the door? Door ladders and spice racks screwed to the inside surface of a door open up a lot of additional storage. They are about 4½ in. deep, so they don't take up too much room (but don't forget to size drawers and shelves accordingly). They are ideal for smaller, frequently used items and are easily adjustable. The wire shelves simply hang off the rack, so they can be moved up or down in a few seconds.

Making the Most of a Laundry Room

■ BY BYRON PAPA

Many builders rely on fancy kitchens, baths, living rooms, and exterior facades to sell their houses. To give my buyers real value, though, I think that I have to put an equal effort into the utilitarian aspects of my houses.

When I'm designing a house, one of my starting points is the laundry. Features such as laundry chutes (photos, facing page) and towel drawers (drawing and photos, p. 117) that extend directly from the laundry into a bathroom lessen wash-day drudgery.

Locate Laundries Centrally

Laundry rooms play a big role in making a house convenient. A larger laundry room could double as a sewing or hobby room, already set up with ventilation and a sink. Some designs I plan to build have the laundry on the second floor. If all the bedrooms are upstairs, it's logical for the laundry room to be upstairs, too. Then I build two towel drawers, one to each second-floor bath.

I place the laundry in a central spot relative to the bathrooms—that is, adjacent to the first-floor bath, which is directly below a second-floor bath. This way, laundry chutes and towel drawers, not to mention plumbing, are easy to place. And I can build a closet off the laundry for a direct-vented water heater. This central location is close to showers and faucets, so there isn't a wait for hot water. I try to put other mechanicals here, such as the electrical panel (photo, p. 114), the central vacuum, and the alarm controls.

Opening the house's primary side door into the laundry is popular, particularly with families that have young children. If you regularly hang clothes outside to dry, the trek to the clothesline is shorter.

The laundry is also a great spot for a closet to store cleaning gear. I install hooks for a heavy-duty mop and broom, and leave room for a vacuum cleaner.

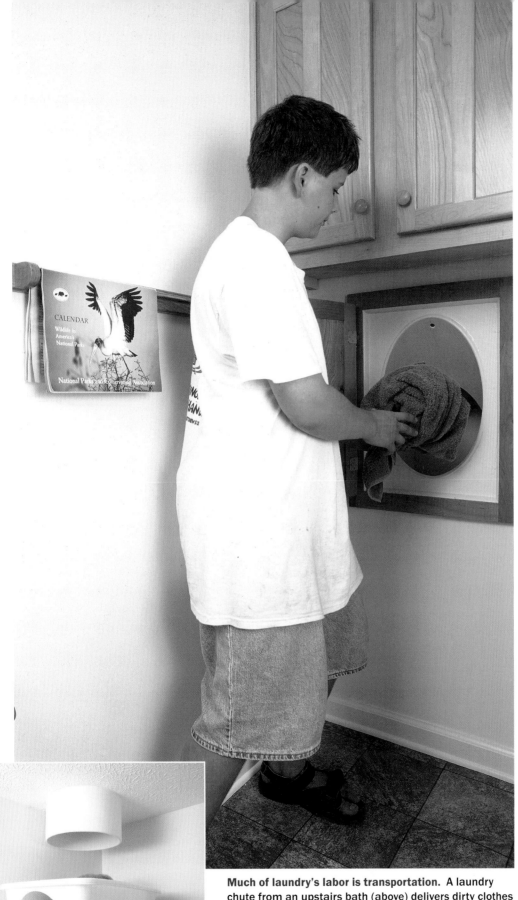

Much of laundry's labor is transportation. A laundry chute from an upstairs bath (above) delivers dirty clothes and towels to a first-floor laundry room (left).

Appliances Go on an Exterior Wall

If the laundry room is the main side entry to the house, it's imperative to locate the appliances and fixtures to maintain an unobstructed path through the room. The laundry room is a workroom where the debris of half-finished tasks may linger. I try to anticipate these scenarios and design around them. For example, I like to include a foldaway ironing board, positioned so that even when it's open, it's out of traffic's way (photo, facing page). I use one by NuTone®. Most foldaway ironing boards fit between 16-in. o. c. studs, so if one isn't in the budget right away, you can easily add it later. Just be sure there's a clear stud bay in the right spot and a nearby electrical outlet.

I position the washer and dryer on an exterior wall so that the dryer vent can be short and straight. In cold climates, though, you might think twice before locating plumbing on outside walls. The washer shouldn't be far from the laundry-chute hampers. Other considerations are a surface for folding laundry and a bar for hanging clothes.

A shelf or cabinet over the washer is great for storing laundry supplies. With the new front-loading washers, which use about half of the hot water that a conventional washer does, you no longer need access from above, and any cabinets or shelves can be built lower. For my next house, I might opt to install a cabinet that has a shelf below it, just above the appliances.

No laundry room of mine is without a utility sink, complete with a sprayer. With faucet, these bargain sinks cost about $90. They are a great spot to clean muddy boots, paint rollers, and small pets. A built-in bench near the sink makes a handy spot to sit and take off muddy boots.

Even in a small house, a laundry can accommodate more than a washer and dryer. In about 80 sq. ft., this laundry packs in a folding table, a towel drawer, a laundry chute, an ironing board and— out of sight—a utility sink.

Fan Helps to Dry Wet Clothes

The moisture level in a laundry room is bound to be high, so it's a good idea to install a vent fan like one that would be found in a bath. I put these fans on built-in timers. I think they're more likely to be used if the homeowner doesn't have to remember to turn off the fan. My favorite fans are made by Panasonic® and NuTone. Panasonic fans are whisper quiet, but NuTone offers more options. I size the fan based on the manufacturer's recommendation for a bathroom of a similar size.

To ensure long-term durability, I vent the fan and dryer through 4-in. sheet-metal ducts. I pop-rivet and seal the joints with mastic or silicone. For energy efficiency, all the bath and laundry fans duct outside through a terminal with a backdraft-preventing door. For the same reason, I vent the dryer through Tamarack Technologies's dryer-vent terminal.

I install a short closet rod between a window and the vent fan for drying delicate garments. With the window cracked, the vent fan moves air through the garment, drying it quickly and taking the moisture out of the house. If this rod can be put over the sink, it makes a perfect location to hang dripping-wet swimsuits.

If ever a room begged for bright fluorescent lighting, the laundry is it. A standard double 4-ft. fluorescent with warm-light bulbs does an excellent job.

Satin Paint for Easy Cleaning

A laundry room sees a lot of action, so all walls and trim should be coated with durable paint. Glossier paints are easier to clean, but they highlight imperfections in the painted surface. Walls look best covered

Sources

Panasonic
1 Panasonic Way
Secaucus, NJ 07094
(800) 222-4213
www.panasonic.com

Energy Federation Inc.
40 Washington St.
Westborough, MA 01581
(800) 379-4121
www.efi.org

Tamarack Technologies
P.O. Box 490
W. Wareham, MA 02576
(800) 222-5932
www.tamtech.com

NuTone, Inc.
4820 Red Bank Rd.
Cincinnati, OH 45227
(888) 336-3948
www.nutone.com

PVC Sewer Pipe Makes a Great Laundry Chute

I build laundry chutes from 14-in.-dia. schedule-40 PVC pipe. In the second-floor bathroom, a door opens to a 12-in. PVC feeder chute that flows into the main vertical chute at a 45° angle (top drawing, facing page). The PVC costs about $100 per house and is machined with saws and routers. Its smooth inside won't snag socks, and the outside can be coated with acrylic paint to match the laundry's ceiling.

The only drawback I've found is that the pipe comes in 20-ft. lengths and that a typical house takes only 7 ft. I buy enough for several houses at once.

The chase for the laundry chute must be designed into the house from the outset. The bath above must line up with the laundry room so that the main chute has a straight drop.

Framing is simple: I lay out the second floor so that the chute fits between two joists, and I nail solid blocking between the joists to box in the chute. Then I build stud walls on the second floor above this box. No additional framing is required on the first floor.

I cut the main chute to length, belt-sand the cut smooth, then round over the bottom with a ³⁄₁₆-in. radius bit. I rabbet the inside of the top of the chute with a bearing-guided router bit to receive a round cap of laminate-faced ¾-in. birch plywood. To let me reach inside to mount the chute, I set the cap last, after the chute is screwed in place.

The chute is raised into the chase, and reaching inside, I screw it to the framing with at least ten #12 by 2½-in. flathead stainless-steel sheet-metal screws. I predrill and countersink all the screw holes. Sooner or later, a kid will climb down any laundry chute, so don't skimp on blocking and screws.

After securing the main chute, I cut a 45° angle on a long piece of 12-in. feeder chute with a jigsaw. Holding this in place against the main chute, I scribe-fit the feeder chute to the main chute, trimming with a belt sander. When I like the fit, I mark the feeder chute's outer edge on the main chute and cut to this line with a jigsaw. The

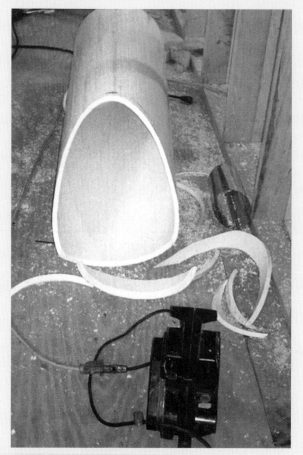

Fitting the feeder to the main chute. A belt sander is the tool to trim the feeder chute to fit with the main chute.

feeder chute now just fits inside the main chute.

Then I mark and cut the feeder chute where it enters the bathroom and sand the cut edges smooth. I screw the feeder chute to horizontal blocking and to the wall framing. I use silicone caulk to seal it to the main chute; a piece of ¾-in. plywood fits around the feeder chute's open end. A cabinet door closes it off.

Any penetration between floors is a potential path for a fire to follow. To block this path, it's important to seal around the chute where it penetrates the floor with a collar of at least 18-ga. steel. Tom O'Loskey, a building official in Connecticut, calls for a smoke detector, linked to the other alarms in the house, to be installed inside laundry chutes at the cap.

in flat paint, but flat paint is hard to clean. I compromise with semigloss for trim and satin on walls. I'm not loyal to a particular paint manufacturer but go along with *Consumer Reports* recommendations. They rate wall paints about every three years.

For easy maintenance, I opt for a high-quality sheet-vinyl floor. Laundries are small rooms that see heavy use. It doesn't pay to skimp on flooring quality here. Again, I base my choice on *Consumer Reports* ratings.

Byron Papa is a building and remodeling contractor in Durham, North Carolina.

Drawer closes into the laundry. With the lid open, towels can be taken from the dryer and stored in the drawer.

And opens into the bath. A deep drawer ensures there's always a clean towel when it's needed.

The Laundry Chute Is Planned from the Outset

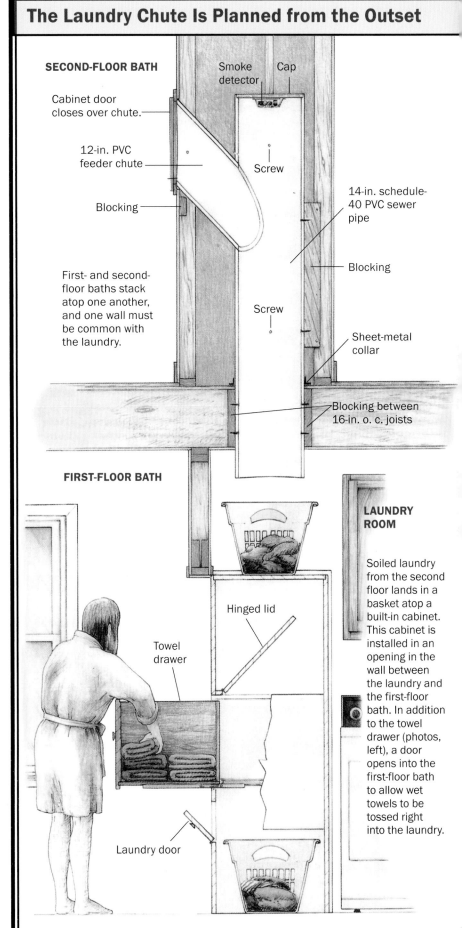

SECOND-FLOOR BATH

Cabinet door closes over chute.

Smoke detector

Cap

12-in. PVC feeder chute

Screw

14-in. schedule-40 PVC sewer pipe

Blocking

Blocking

First- and second-floor baths stack atop one another, and one wall must be common with the laundry.

Screw

Sheet-metal collar

Blocking between 16-in. o. c. joists

FIRST-FLOOR BATH

LAUNDRY ROOM

Hinged lid

Towel drawer

Laundry door

Soiled laundry from the second floor lands in a basket atop a built-in cabinet. This cabinet is installed in an opening in the wall between the laundry and the first-floor bath. In addition to the towel drawer (photos, left), a door opens into the first-floor bath to allow wet towels to be tossed right into the laundry.

Hide that Ugly Refrigerator

■ BY REX ALEXANDER

A run of handmade kitchen cabinets bumping up against the stark metal surface of a refrigerator makes my skin crawl. That's why I try to camouflage this homely appliance: To make it more appealing to the eye and more in harmony with the rest of the kitchen, I build it into the cabinetry. It's a relatively inexpensive way to add charm and sophistication to any kitchen.

Building in Gives the Kitchen a Finished Look

Generally, what distinguishes an ordinary built-in refrigerator from a true built-in, such as a Sub-Zero®, is the depth of the unit from front to back. Conventional refrigerators are 32 in. deep. True built-ins vary in depth from 23½ in. to 25⅜ in., a real plus in a small kitchen where the appliance needs to be flush with the cabinetry.

Because many of my customers still use conventional refrigerators (normally 32 in. square by 65 in. tall), I developed my own method of building in these appliances. This amounts to side panels and a cabinet above the refrigerator. The side panels run the height of the refrigerator and form the sides of the upper cabinet (photo, facing page). If the refrigerator doesn't come with doors that accept panels, I leave the doors as they come from the factory.

First, Notice How the Door Is Hinged

Before building in a conventional refrigerator, I have to decide dimensions, clearances, and the operation of the door in relation to the cabinet. I first look at how the refrigerator door is hinged.

I always size the width of the upper cabinet to coincide with the width of the refrigerator opening. The doors of many units open and stay parallel to the side. With a ¾-in. to 1-in. spacing around the unit, my cabinet can be flush with the door, although the handle will stick out. The newer doors that hold gallon-milk containers open with the thickness of the door extending beyond the side. For these, I hold open the door and set the distance between refrigerator and side panel and the depth of the side panel accordingly. That way, the door can open fully.

Minimize the impact of a metal box in a roomful of wooden ones. A basic refrigerator enclosure that matches the rest of the kitchen cabinetry adds a lot to the looks of a kitchen. The cabinet over the refrigerator provides storage.

Once I've established the depth of the refrigerator cabinet, I complete a construction drawing for both the cabinet above the refrigerator and the side panels. I like to keep the clearance around the refrigerator as narrow as possible. So I measure the width of the icebox and add 1½ in. to 2 in. to the width of the cabinet that goes on top. Manufacturers recommend a ¾-in. to 1-in. clearance around the sides, top, and back. Newer refrigerators have cooling coils on the bottom, so there's no need for much ventilation.

I've built cabinets over refrigerators that run the full depth of the icebox—anywhere from 28 in. to 32 in. (photo, right). However, I've found that keeping the depth at 24 in. is more convenient. At that height, it's easier to retrieve things from a shallow space. A shallow cabinet also gives the refrigerator breathing space.

Construction Is as Simple as Building a Box

No special construction techniques are involved in building the side panels or the upper cabinet. Both the style of the built-in and the way the pieces fasten together depend on the style of the other cabinetry.

Depending on how much of the sides show, the sides can display various raised or flat panels, or I can use plain, edged plywood. Cabinet doors above the refrigerator typically match the rest of the kitchen's cabinet doors. If the top of the cabinet is still a distance below the ceiling, I usually trim the top of the refrigerator built-in with some type of molding. Otherwise, I match the cabinet molding to the crown molding around the rest of the kitchen.

I use ¾-in. stock for the cabinet and side panels. I usually tack ¼-in. plywood or melamine to the back of the upper cabinet. Side panels can be built with rails and stiles applied to plywood to simulate rail-and-stile construction or with plywood finished to match the kitchen.

Refrigerator cabinetry can match any style of kitchen. Even the most sophisticated kitchen can be improved with the addition of a refrigerator enclosure.

Finally, Put It All Together and Fasten in Place

With the side panels, top cabinet, and moldings completed, I usually move the refrigerator in place and build around it. I level and scribe-fit the side panels to the wall and floor, then shim the top cabinet using the top of the refrigerator as a level base. Next, I remove the refrigerator and, from the inside of the cabinet, screw the top cabinet to the top of the side panels. I finish by applying the molding to the top of the whole unit.

I put a large piece of laminate on the floor, set the refrigerator on top of it and then slide the appliance into place. After the refrigerator is installed, I tilt it up and pull out the laminate.

Rex Alexander, a frequent contributor to Fine Homebuilding *and* Fine Woodworking, *builds cabinetry and furniture from his home shop in Brethren, Michigan.*

Faux Fridge Front

■ BY MIKE GUERTIN

From the outset, our kitchen plans included a refrigerator that looked like part of the cabinetry (photo, right). After looking at the price tags on several models designed to accommodate cabinet-matching door panels and built-in dimensions, I had to find a better way.

Solving the depth problem was easy because I was building the cabinets myself; I just framed the refrigerator cabinet deeper to accommodate a standard refrigerator. I avoided, however, thinking about the panel-mounting problem until the last minute. My wife vetoed the idea of screwing or gluing the panels to the refrigerator doors. Defacing a brand-new appliance just wasn't an option. Then, in a moment of frustration and desperation, came a little inspiration.

The Wood Doors Hang on Aluminum Channels

My plan involved metal channels shaped like a J. The channels are screwed to the backs of the wood panels and wrap around the refrigerator door (drawing, p. 123).

I started with some leftover flat aluminum 0.032-in. gutter stock finished black.

Find the fridge. At a glance, it's tough to tell where the cabinets end and the refrigerator begins, except when you're hungry.

Aluminum channel is the key. Bent into a J-shape, aluminum channel is screwed to the back of the cabinet panel. A gap in the top channel accommodates the refrigerator-door hinge.

I figured the dark color would tend to draw less visual attention to the edges of the refrigerator door than a lighter color. I own a brake for bending aluminum, but for those who don't, finding someone to fabricate the channel shouldn't be a problem.

First, I measured the distance from the gasket to the edge of the door and subtracted ⅛ in. so that the aluminum wouldn't interfere with the gasket. That was my first bend (drawing, facing page).

I made my second bend at 2½ in. (¼ in. more than the thickness of the door) to wrap around the door. The extra width makes it easier to slide the channels over the doors. On the leg that attached to the panel, I added a gentle S-bend. That bend accommodates the screws and the raised panel.

After testing one channel on the edge of the refrigerator door, I decided to overbend each 90° angle by 1° or 2° for a snug fit. I cut three legs for each door, one for each side and the top. The top channels are 2 in. short to avoid the refrigerator-door hinges (photo, above). Because the side channels extend to the bottom of each panel, I didn't put a channel on the underside of the doors.

When I had the wood panels made, I oversized them ¼ in. in both directions. When I screwed the channels to each door panel, the side channels ended up a little farther apart than the actual width of the doors. That extra space kept me from having to force the panels over the doors.

Remove the Door for Easy Installation

Before I could install the panels, I had to remove the doors from the refrigerator. Otherwise the hinge brackets were in the way. With the doors removed, the wood panels slid easily over the refrigerator doors (photo, below).

I rehung the doors on their hinges, adjusted them in the opening, and pushed the refrigerator back into its cabinet. It only took about 20 minutes to remove the refrigerator and freezer doors, slip the panels on, and rehang them, not even long enough to worry about melting my Ben & Jerry's®.

Held in place by gravity and friction. With the door removed, the panel slides easily over the door.

Cross Section of Panel Attachment

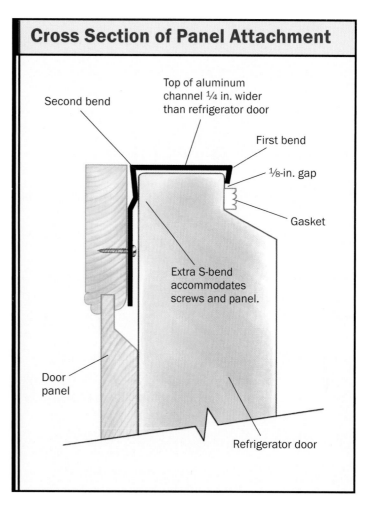

Second bend

Top of aluminum channel ¼ in. wider than refrigerator door

First bend

⅛-in. gap

Gasket

Extra S-bend accommodates screws and panel.

Door panel

Refrigerator door

The only drawback to the system was that the hinges weren't meant to accommodate the extra thickness of the clad door, so we keep the fridge sticking out slightly into the room to let the doors open fully. Friction and gravity keep the door panels in place. If my wife ever decides to redecorate the kitchen, I can easily slip the wood panels off and give the refrigerator a fresh white look.

Mike Guertin is a custom home builder/remodeler in East Greenwich, Rhode Island. He is the author of Roofing with Asphalt Shingles (The Taunton Press, Inc., 2002), and co-author of Precision Framing (The Taunton Press, Inc., 2001).

Home Storage

■ BY JOANNE KELLAR BOUKNIGHT

Nothing separates a great home from a good one like plentiful, thoughtfully designed storage. Good storage not only holds our worldly goods, it also lets us live more organized lives. Just consider how much better it would be to take the time spent looking for misplaced stuff and to spend it on family.

There's certainly no single way to achieve storage Zen, but some basic guidelines can help to clear the path to good storage. The first step is always to get rid of everything you don't really need or love. This step makes room to design storage specifically for those objects you can't live without, and then some.

Ideally, storage-space design takes into account the habits of the people whose lives it will organize because what's right for one family won't necessarily be right for another. For builders and remodelers, this means talking to homeowners about their habits and priorities. For your own house, the first

Coming or going? This back hall has a handy space to drop off or pick up anything. Drawers, cabinets, cubbies, and a wide window shelf offer plenty of storage. Even the kids' artwork is preserved.

Top-hinged picture frames preserve children's artwork.

Counter is a landing zone for incoming groceries and outgoing parcels.

Cubbies provide quick access to some items while drawers hide clutter.

Stairways and Entryways

An efficient house, no matter how small, designates space for an entryway: a space for hanging up coats and hats, for stashing a wet umbrella, for taking off boots. It has a place for plopping down mail and groceries, and a row of hooks for backpacks. A well-designed entryway helps to organize a family's daily life, eliminating frantic morning searches for keys and shoes.

Even in a small house, carving out a bit of room for better organizing inbound and outbound traffic (and all the stuff that goes along for the ride) is space well spent because it reduces dirt and clutter throughout the rest of the house. A hall desk, a cantilevered shelf, or a niche carved into a thick wall each could work. And no one ever regrets making space for a full mudroom.

Stairways represent an often-missed storage opportunity. Display shelves and bookshelves can line the stairs, while the area underneath the stairs often affords space for another closet or built-in shelves and drawers for holding out-of-season clothing, linens, footwear, or opening gear.

Mudrooms keep outdoor gear organized. Benches, pegs, open shelves, closed drawers, and a closet keep clothes and equipment sorted after sports activities.

Who needs a library? Both sides of this open stairway are faced with bookshelves and display shelves.

trick to efficient storage is to look at how you store things and list what works, what doesn't, and why. Can you find your keys every morning? Can you find once-a-year decorations quickly when the time comes? Are you tripping over boots and briefcases near the front door?

A mudroom with an adjacent laundry and an extra bathroom may be the perfect antidote to a family's entryway chaos, and a walk-in closet or dressing room makes better use of space than a huge bedroom filled with furniture. And even a small pantry can allow a kitchen to function more efficiently without clutter.

Before settling on how to store your possessions, consider how often you'll use them. Daily use calls for different strategies than annual use. Car keys, newspapers, backpacks, and bath towels should be within arm's reach but shouldn't clutter countertops. Seasonal clothing or holiday ornaments are dormant for months at a time; so although they should be accessible, they needn't be at hand.

Storage can be improved by the smallest of changes or by the biggest of commitments, from the addition of a few shelves to the addition of an entire mudroom, depending on the time, money, space, and thought budgeted for the project. An existing home can become more efficient simply with the addition of a few hooks and pegs. Cabinets and closets can be reconfigured, and shelves can be added to unused wall space. Square footage may not be available for additional storage, but you're bound to find space somewhere that can be better designed. Even small changes can make a big difference.

Joanne Kellar Bouknight is the author of The Kitchen Idea Book *and* The Home Storage Idea book, *both published by The Taunton Press, Inc. She lives in Cos Cob, Connecticut.*

Kitchens

Some cooks prefer to keep cooking tools within easy reach on the countertop or in open shelves, while others like having a drawer to hide everything. Is the kitchen a compact space for one cook, or is it a large, common room designed to accommodate kids doing homework and several guests while two cooks prepare dinner? The storage needs of these kitchens differ, but a well-designed selection of closed cabinetry, open shelving, and generous pantry storage will always help a handsome kitchen to look neat and operate smoothly.

Cutting corners. An unusual corner drawer makes good use of a usually wasted space, and a bifold door eases access to the base cabinet.

Who gets the corner office?
A sliver of desk accommodates the basics—commonly used files, extra drawers, and a writing surface—while a bulletin board/chalkboard combo offers space for phone messages and grocery lists.

Message center by phone is accessible from seated or standing position.

A small writing desk makes it easy to pay bills, check schedules, or write notes to a teacher.

Family papers stay within reach but out of the way.

(Continued next page)

A fold-out pantry. The narrow double doors on this pantry provide the perfect space for canned goods and condiments. Small plastic wire baskets hold soup mixes and bags of flour; a larger one holds potatoes.

The delight is in the details. Drawer details differ according to size. Shallow drawers have flush fronts, while deeper drawers support frame-and-panel designs. A recessed towel bar keeps dish towels out of the way but within reach.

Built-ins where they belong. An undercounter container keeps vegetable trimmings off the work surface while they await a trip to the compost heap, and built-in knife and utensil storage keeps food-preparation tools exactly where they're needed.

Bedrooms

Like the kitchen pantry that is required to contain ten brands each of cereal, cookies, and soup, today's closet has to accommodate a colossal variety of clothing.

The heart of storage in bedrooms is the closet, be it the standard 2-ft.-deep wall closet or a room-size walk-in. In either event, closet space goes a lot further if properly outfitted with shelving, books, rods, and cubbies designed to fit the clothes that they store (see "Outfitting a Clothes Closet," pp. 4–14).

Drawer storage is best for small items such as lingerie, underwear, socks, swimsuits, and exercise clothes. Sweaters and knit shirts can go in drawers, but they are easier to see and access on shelves. It's better to space shelves closely for folded items than to build a leaning tower of sweaters. Count and measure everything you'll be keeping, and build in extra space for adding clothes.

Most clothes are best stored hanging, so the more poles the better: both single poles for coats and dresses, and double poles for shorter items. Short spans for each type of clothing work better than mixing everything on one long pole. Robes, towels, work shirts, belts, and scarves work on pegs.

Everything in its place. This dressing room has every kind of storage a couple needs. Closet rods accommodate clothing of all lengths, and adjustable shelves waste no space. At the far end, a sewing center means repairs can be done where the clothes are.

His and hers closets. The closet walls stop short of the ceiling to admit natural light. Positioning the closets at the entry to the bedroom builds in more privacy. The adjoining study keeps work-related clutter out of the bedroom.

(Continued next page)

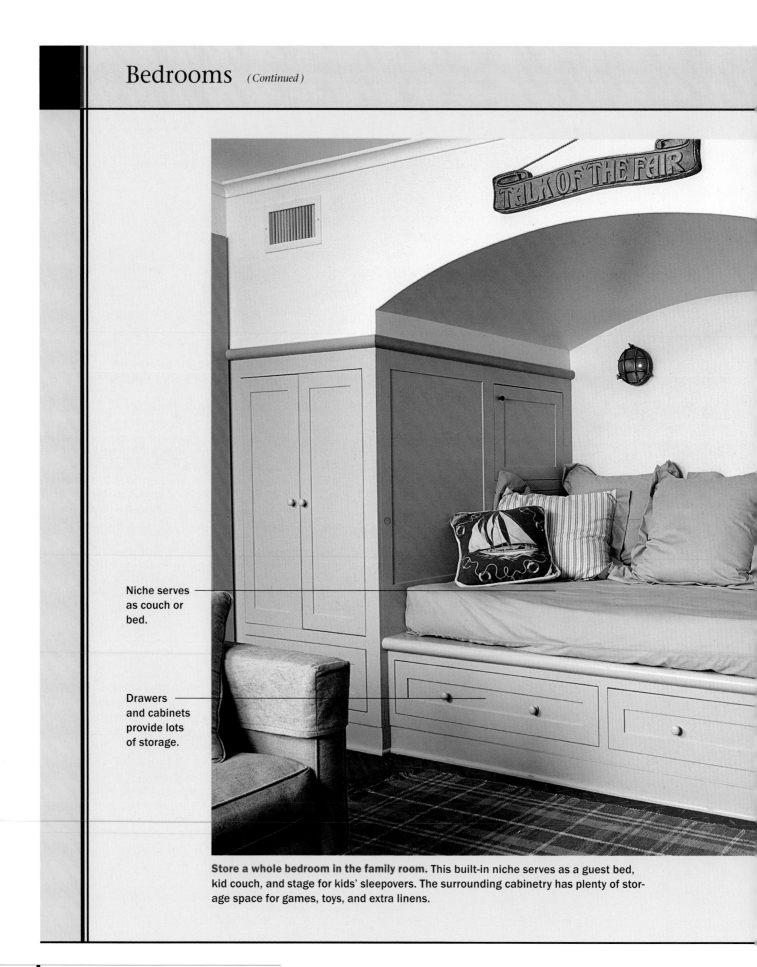

Niche serves as couch or bed.

Drawers and cabinets provide lots of storage.

Store a whole bedroom in the family room. This built-in niche serves as a guest bed, kid couch, and stage for kids' sleepovers. The surrounding cabinetry has plenty of storage space for games, toys, and extra linens.

Working at playing. A closet-top loft provides a getaway space and keeps stuffed animals off the floor. Clothes are all within view and are easy to retrieve and return. Resembling a playhouse, the closet also works to define the desk to the left as a study area.

A light in the attic. A skylight that creates headroom and light in this attic closet makes the perfect niche for a built-in bureau. Clothes hang within easy reach on each side.

Storage
Ideas

Although a clever storage idea, drawers in stair risers are also potentially dangerous—leaving a drawer open could trip someone. These attic stairs, however, get little traffic, and the drawers feature self-closing slides (wedged open for the photo).

A shoe-storage bin rolls out on full-extension drawer slides beneath this stair.

A bookcase devoted to cookbooks swings open over a granite counter, revealing more storage for kitchen items.

Inspired by **Norwegian architecture** and designed some 50 years ago by Edwin Lundie, this Minnesota cabin sports a modest kitchen that hides behind a handsome pair of bifold doors.

Finishing Touches: Hang 'Em High

An overhead custom storage rack keeps pots and pans at the cook's fingertips without cluttering up the workspace. Because it's installed over the island, the homeowners avoid banging their heads against a pot or pan accidentally.

This wall-mounted rack designed and built by Charles Miller looks good and reduces clutter.

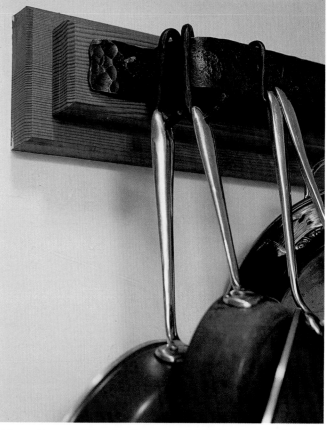

This custom bracket by Valerie Walsh neatly attaches under a skylight and cantilevers the pot rack, allowing for quick access within arm's reach.

A forged iron and Douglas fir pot rack highlights both craft and materials while providing efficient storage for bulky cookware.

Built-In Seating

A corner for some tea and a book. In an alcove next to the kitchen, this seat can rest the bones of a weary cook or delay the ravenous horde.

A spot to cool your heels.
This couch invites a good read
beneath plenty of bookcases
and a decorative screen.

**Vertical-grain fir gives a
warm glow** to this window
seat. Designed by Stephen
Bobbit, this seat is built
into a dormer in a house
near Puget Sound.

Her master's voice, indeed. This bed and cabinet unit becomes a couch by shifting some cushions and the occasional couch potato.

Safe in the garden. Hand-painted trellises surround this seat that faces the Green Mountains.

A mahogany nook beneath the stairs is an unexpected oasis. When Robert Prutting remodeled his house, he built this quiet spot under an Arts and Crafts staircase.

A place in the sun. Designed and built by the South Mountain Company, Inc., this built-in couch was made from salvaged cypress.

Creative Storage in a Small House

Specialized storage for a cook's kitchen.
Copper and glass cabinet fronts add a different
texture to this modern kitchen (right).
Designed by the owners, custom airtight containers (below) store dry goods within easy
reach of the work area.

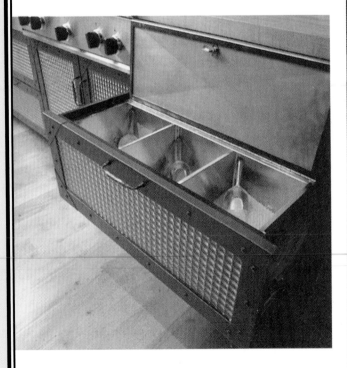

To maintain the small size of the house he was designing on the coast of Washington state, architect Barry Gehl thought in terms of building a boat. After all, boat designers know the value of smart storage space and fixtures that perform double duty.

He designed a screen wall (photos, p. 147) to separate the main bedroom from the living area. Made of welded tube steel, plywood, and reinforced plaster, the pivoting wall section/bookcase weighs about 700 lb. Carpenters Larry Gladstone and Randy Kent solved the problem of hardware: They used an old truck axle that's bolted to a concrete pad below. The wheel bearings carry the weight. Gehl used the thick walls for more than a barrier; he filled them on the bedroom side with cabinets and bookshelves (bottom photo, p. 146).

Adding to the creative mix, the owners designed the main walk-in closet (top photo, p. 146) and had a hand in designing the kitchen (photos, below). The kitchen-cabinet fronts, acid-washed and riveted copper frames that surround textured-glass panels, were crafted by Doug French and seem to be lighted from within by the contents of the cabinets.

Use the vertical to maximize space. By designing stacked drawers and angled shelves, the owners were able to squeeze the most room from a narrow closet.

Wall section swings to reveal the bedroom and bookcases. To separate the master bedroom (left) from the living room (above), the architect designed a thick screen wall; a pivoting middle section (above left) serves as a door and bookcase.

Entertainment Centers

Hiding the television until you need it. A renovated family room over the garage features a concealed projector and a screen that drops down out of the ceiling on a remote-actuated cable lift.

Going vertical saves room.
John Marckworth built this
recessed unit for architect
Frank Karreman's house
near Seattle. The maple
cabinets hold books along
the sides; vertical pullouts
hold CDs and videos.

Making a case for curves. Handed a request to
make a home-style jukebox for a Minneapolis
house, Jon Frost took a sculptural approach and
built a sleek cabinet from mahogany-veneered
wheatstraw particleboard.

Deep in the heart of home.
Designed by Dick Clark Architects and built by Bill Johnson of Austin, Texas, this alder assembly hides larger speakers behind the lower grilles and complements a renovated 1960s house design.

The cool hearth. In the Denver home of a horticulturist, a video screen recessed into the wall is complemented by inlaid frosted glass backlit by fluorescent light.

Simplicity does the job. A stand-alone oak cabinet with clean lines and rustic door pulls camouflages the television when it's not being watched.

Under the watchful eye. In a new addition, Austin, Texas, architect Walter Toole designed a structure that contains a television on the side shown; the opposite side holds audio equipment. The copper-rimmed portal adds an unusual touch.

Towering media. Mark Chestnutt of Stephen Blatt Architects, Portland, Maine, designed this structure to provide office space on its back side, hold audio hardware on its shelves, and host a reading nook at the top.

CREDITS

The articles in this book appeared in the following issues of *Fine Homebuilding*.

p. iii: Photo by Roe A. Osborn, courtesy of *Fine Homebuilding,* © The Taunton Press, Inc.

p. iv: (left) Photo by Kevin Ireton, courtesy of *Fine Homebuilding,* © The Taunton Press, Inc.; (right) Photo by Scott Gibson, courtesy of *Fine Homebuilding,* © The Taunton Press, Inc.

p. v: (left) Photo by Scott Gibson, courtesy of *Fine Homebuilding,* © The Taunton Press, Inc.; (center) Photo by Roe A. Osborn, courtesy of *Fine Homebuilding,* © The Taunton Press, Inc.; (right) Photo © George Anns

p. 4: Outfitting a Clothes Closet by Gary Katz, issue 124. Photos by Roe A. Osborn, courtesy of *Fine Homebuilding,* © The Taunton Press, Inc.

p. 15: Simple Closet Wardrobe by Jim Tolpin, issue 78. Photos © Jim Tolpin; Illustrations © Jim Tolpin

p. 22: Custom Closet Wardrobe by Philip S. Sollman, issue 83. (pp. 22–23) Photos by Kevin Ireton, courtesy of *Fine Homebuilding,* © The Taunton Press, Inc.; (pp. 25–28) Photos © Philip S. Sollman; Illustration by Michael Hiotakis

p. 30: Building a Fold-Down Bed by Patrick Camus, issue 138. Photos by Charles Bickford, courtesy of *Fine Homebuilding,* © The Taunton Press, Inc.; Illustrations by Bob LaPointe

p. 34: Fanciful Built-In Beds by Jean Steinbrecher, issue 118. Photos by Charles Miller, courtesy of *Fine Homebuilding,* © The Taunton Press, Inc.; Illustration by Bob LaPointe

p. 40: Bed Alcove by Tony Simmonds, issue 76. (pp. 40–41) Photo by Charles Miller, courtesy of *Fine Homebuilding,* © The Taunton Press, Inc.; (pp. 44–47) Photos by Tony Simmonds; Illustrations by Bob LaPointe

p. 48: A Bookcase That Breaks the Rules by Gary M. Katz, issue 154. Photos by Roe A. Osborn, courtesy of *Fine Homebuilding,* © The Taunton Press, Inc., except p. 52 Photos © Dean Della Ventura; Illustrations by Bob LaPointe

p. 57: Designing Built-Ins by Louis Mackall, issue 111. Photos by Kevin Ireton, courtesy of *Fine Homebuilding,* © The Taunton Press, Inc.

p. 66: A Pair of Built-In Hutches by Kevin Luddy, issue 125. Photos by Roe A. Osborn, courtesy of *Fine Homebuilding,* © The Taunton Press, Inc.

p. 72: Building a Lazy-Susan Cabinet by Rex Alexander, issue 133. Photos by Scott Gibson, courtesy of Fine Homebuilding, © The Taunton Press, Inc., except p. 75 (top) Photo © Al Amstutz

p. 80: A Built-In Hardwood Hutch by Stephen Winchester, issue 85. (p. 81 bottom, p. 87) Photos © Stephen Winchester; (p. 81 top, p. 86) Photos by Rich Ziegner, courtesy of *Fine Homebuilding,* © The Taunton Press, Inc.; Illustrations by Bob Goodfellow

p. 88: Designing and Building an Entertainment Center by Brian Wormington, issue 120. Photos by Andy Engel, courtesy of *Fine Homebuilding,* © The Taunton Press, Inc., except p. 93 (inset) Photo by Scott Phillips, courtesy of *Fine Homebuilding,* © The Taunton Press, Inc.; Illustrations by Vince Babak

p. 94: Building a Fireplace Room Divider by Alexander Brennen, issue 90. Photos by Charles Miller, courtesy of *Fine Homebuilding,* © The Taunton Press, Inc.; Illustrations by Bob Goodfellow

p. 99: Improving Kitchen-Cabinet Storage by Sven Hanson, issue 161. Photos by Dan Morrison, courtesy of *Fine Homebuilding,* © The Taunton Press, Inc.; Illustrations by Dan Thornton

p. 106: Updating the Kitchen Pantry by David Getts, issue 121. Photos by Scott Gibson, courtesy of *Fine Homebuilding,* © The Taunton Press, Inc.

p. 112: Making the Most of a Laundry Room by Byron Papa, issue 121. Photos by Andy Engel, courtesy of *Fine Homebuilding,* © The Taunton Press, Inc., except p. 116 Photo by Byron Papa; Illustrations by Christopher Clapp

p. 118: Hide that Ugly Refrigerator by Rex Alexander, issue 108. Photos by Steve Culpepper, courtesy of *Fine Homebuilding,* © The Taunton Press, Inc.

p. 121: Faux Fridge Front by Mike Guertin, issue 127. Photos by Roe A. Osborn, courtesy of *Fine Homebuilding,* © The Taunton Press, Inc.; Illustration by Paul Perreault

p. 124: Home Storage by Joanne Kellar Bouknight, issue 149. (pp. 124–125, 132) Photos © Randy O'Rourke; (p. 126) Photo © Robert Benson; (pp. 127, 129, 130 top and bottom right, 131 bottom) Photos © David Duncan Livingston; (pp. 128, 130 bottom left, 131 top) Photos © Robert Perron; (p. 133 top) Photo © Charles Register; (p. 133 bottom) Photo © Derrill Bazzy

p. 134: Storage Ideas, issue 85. (p. 134) Designed by, built by and photo © John Hermannsson; (p. 135) Built by and photo © George Anns; (p. 136) Photos © Brian Vanden Brink, Photographer 2004; (p. 137) Photos by Scott Gibson, courtesy of *Fine Homebuilding,* © The Taunton Press, Inc.

p. 138: Finishing Touches: Hang 'Em High, issue 81. Photos by Charles Miller, courtesy of Fine Homebuilding, © The Taunton Press, Inc.; (p. 138) Charles Montooth design, built by Ray Martinez

p. 140: Finishing Touches: Built-In Seating, issue 113. (p. 140 bottom) Photo © Robert Perron; (pp. 140-141 top) Design by Brian Olfe, Photo © Dan Olfe; (p. 141) Photo © Stephen Rosen; (p. 142 top) Design by Robert Knight, Photo © Brian Vanden Brink, Photographer 2004; (p. 142 bottom) Painting by Denise Welch-May and Helen Foster, Photo © Robert Perron; (p. 143 top) Photo by Charles Bickford, courtesy of *Fine Homebuilding,* © The Taunton Press, Inc.; (p. 143 bottom) Photo © Derrill Bazzy

p. 144: Finishing Touches: Creative Storage in a Small House, issue 121. Photos © Chris Eden

p. 48: Finishing Touches: Entertainment Centers, issue 155. (p. 148) Design by Custom Electronics, Portland, Maine, Photo © Brian Vanden Brink, Photographer 2004; (p. 149 top) Photo © John Marckworth; (pp. 149 bottom) Photo © Norbert Marklin; (p. 150 top and bottom left) Photo © Paul Bardagiy/Through the Lens Mgmt; (p. 150 center) Design by Joseph Moore, Photo © Povy Kendall Atchison; (p. 150 right and inset) Designed and built by Anatole Burkin, Photo by Anatole Burkin, courtesy of *Fine Homebuilding,* © The Taunton Press, Inc.; (p. 151) Photo © Brian Vanden Brink, Photographer 2004

INDEX